Software Development for Embedded Multi-core Systems

Software Development for Embedded Multi-core Systems

A Practical Guide Using Embedded Intel® Architecture

Max Domeika

AMSTERDAM • BOSTON • HEIDELBERG • LONDON
NEW YORK • OXFORD • PARIS • SAN DIEGO
SAN FRANCISCO • SINGAPORE • SYDNEY • TOKYO

Newnes is an imprint of Elsevier

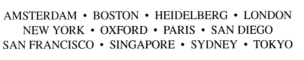

Cover image by iStockphoto
Newnes is an imprint of Elsevier
30 Corporate Drive, Suite 400, Burlington, MA 01803, USA
Linacre House, Jordan Hill, Oxford OX2 8DP, UK

Library of Congress Cataloging-in-Publication Data
Domeika, Max.
   Software development for embedded multi-core systems : a practical guide using embedded
   Intel architecture / Max Domeika.
       p. cm.
   ISBN 978-0-7506-8539-9
   1. Multiprocessors.   2. Embedded computer systems.   3. Electronic data processing—
Distributed processing.   4. Computer software—Development. I. Title.
   QA76.5.D638 2008
   004'.35—dc22   2008006618

British Library Cataloguing-in-Publication Data
A catalogue record for this book is available from the British Library.

For information on all Newnes publications
visit our Web site at www.books.elsevier.com

ISBN: 978-0-7506-8539-9

Transferred to Digital Printing in 2011

Contents

Preface

At the Fall 2006 Embedded Systems Conference, I was asked by Tiffany Gasbarrini, Acquisitions Editor of Elsevier Technology and Books if I would be interested in writing a book on embedded multi-core. I had just delivered a talk at the conference entitled, "Development and Optimization Techniques for Multi-core SMP" and had given other talks at previous ESCs as well as writing articles on a wide variety of software topics. Write a book – this is certainly a much larger commitment than a presentation or technical article. Needless to say, I accepted the offer and the result is the book that you, the reader, are holding in your hands. My sincere hope is that you will find value in the following pages.

Why This Book?

Embedded multi-core software development is the grand theme of this book and certainly played the largest role during content development. That said, the advent of multi-core is not occurring in a vacuum; the embedded landscape is changing as other technologies intermingle and create new opportunities. For example, the intermingling of multi-core and virtualization enable the running of multiple operating systems on one system at the same time and the ability for each operating system to potentially have full access to all processor cores with minimal drop off in performance. The increase in the number of transistors available in a given processor package is leading to integration the likes of which have not been seen previously; converged architectures and low power multi-core processors combining cores of different functionality are increasing in number. It is important to start thinking now about what future opportunities exist as technology evolves. For this reason, this book also covers emerging trends in the embedded market segments outside of pure multi-core processors.

When approaching topics, I am a believer in fundamentals. There are two reasons. First, it is very difficult to understand advanced topics without having a firm grounding in the basics. Second, advanced topics apply to decreasing numbers of people. I was at an instrumentation device company discussing multi-core development tools and the topic turned to 8-bit code optimization. I mentioned a processor issue termed partial register stalls and then found myself discussing in detail how the problem occurs and the innermost workings of the cause inside the processor (register renaming to eliminate false dependencies, lack of hardware mechanisms to track renamed values contained in different partial registers). I then realized while the person to whom I was discussing was thoroughly interested, the rest of the people in the room were lost and no longer paying attention. It would have been better to say that partial register stalls could be an issue in 8-bit code. Details on the problem can be found in the optimization guide.

My book will therefore tend to focus on fundamentals and the old KISS[1] principle:

- What are the high level details of X?

- What is the process for performing Y?

Thanks. Now show me a step-by-step example to apply the knowledge that I can reapply to my particular development problem.

That is the simple formula for this book:

1. Provide sufficient information, no more and no less.

2. Frame the information within a process for applying the information.

3. Discuss a case study that provides practical step-by-step instructions to help with your embedded multi-core projects.

Intended Audience

The intended audience includes employees at companies working in the embedded market segments who are grappling with how to take advantage of multi-core processors for their respective products. The intended audience is predominately embedded software development engineers; however, the information is approachable enough for less day-to-day technical embedded engineers such as those in marketing and management.

---

[1] KISS = Keep It Simple, Stupid.

Readers of all experience and technical levels should derive the following benefits from the information in this book:

- A broad understanding of multi-core processors and the challenges and opportunities faced in the embedded market segments.

- A comprehensive glossary of relevant multi-core and architecture terms.

Technical engineers should derive the following additional benefits:

- A good understanding of the optimization process of single processors and multi-core processors.

- Detailed case studies showing practical step-by-step advice on how to leverage multi-core processors for your embedded applications.

- References to more detailed documentation for leveraging multi-core processors specific to the task at hand. For example, if I were doing a virtualization project, what are the steps and what specific manuals do I need for the detailed information?

The book focuses on practical advice at the expense of theoretical knowledge. This means that if a large amount of theoretical knowledge is required to discuss an area or a large number of facts are needed then this book will provide a brief discussion of the area and provide references to the books that provide more detailed knowledge. This book strives to cover the key material that will get developers to the root of the problem, which is taking advantage of multi-core processors.

Acknowledgments

There are many individuals to acknowledge. First, I'd like to thank Rachel Roumeliotis for her work as editor.

I also need to acknowledge and thank the following contributors to this work:

- Jamel Tayeb for authoring Chapter 9 – Virtualization and Partitioning. Your expertise on partitioning is very much appreciated.

- Arun Raghunath for authoring Chapter 8 – Case Study: Functional Decomposition. Thank you for figuring out how to perform flow pinning and the detailed analysis performed using Intel® Thread Checker. Thanks also to Shwetha Doss for contributions to the chapter.

- Markus Levy, Shay Gal-On, and Jeff Biers for input on the benchmark section of Chapter 3.

- Lori Matassa for contributions to big endian and little endian issues and OS migration challenges in Chapter 4.

- Clay Breshears for his contribution of the tools overview in Chapter 4.

- Harry Singh for co-writing the MySQL case study that appears in Chapter 5.

- Bob Chesebrough for his contribution on the *Usability* section in Chapter 5.

- Lerie Kane for her contributions to Chapter 6.

- Rajshree Chabukswar for her contributions of miscellaneous power utilization techniques appearing in Chapter 10.

- Rob Mueller for his contributions of embedded debugging in Chapter 10.

- Lee Van Gundy for help in proofreading, his many suggestions to make the reading more understandable, and for the BLTK case study.

- Charles Roberson and Shay Gal-On for a detailed technical review of several chapters.

- David Kreitzer, David Kanter, Jeff Meisel, Kerry Johnson, and Stephen Blair-chappell for review and input on various subsections of the book.

Thank you, Joe Wolf, for supporting my work on this project. It has been a pleasure working on your team for the past 4 years.

This book is in large part a representation of my experiences over the past 20 years in the industry so I would be remiss to not acknowledge and thank my mentors throughout my career – Dr. Jerry Kerrick, Mark DeVries, Dr. Edward Page, Dr. Gene Tagliarini, Dr. Mark Smotherman, and Andy Glew.

I especially appreciate the patience, support, and love of my wife, Michelle, and my kids, James, Caleb, and Max Jr. I owe them a vacation somewhere after allowing me the sacrifice of my time while writing many nights and many weekends.

Disclaimer

"*" This symbol is used to repersent "Other names and brands may be claimed as the property of others."

The author is not speaking for Intel Corporation. This book represents the opinions of author.

Performance tests and ratings are measured using specific computer systems and/or components and reflect the approximate performance of Intel products as measured by those tests. Any difference in system hardware or software design or configuration may affect actual performance. Buyers should consult other sources of information to evaluate the performance of systems or components they are considering purchasing. For more information on performance tests and on the performance of Intel products, visit Intel Performance Benchmark Limitations.

## CHAPTER 1

Introduction

The proceeding conversation is a characterization of many discussions I've had with engineers over the past couple of years as I've attempted to communicate the value of multi-core processors and the tools that enable them. This conversation also serves as motivation for the rest of this chapter.

A software engineer at a print imaging company asked me, "What can customers do with quad-core processors?" At first I grappled with the question thinking to a time where I did not have an answer. "I don't know," was my first impulse, but I held that comment to myself. I quickly collected my thoughts and recalled a time when I sought an answer to this very question:

- Multiple processors have been available on computer systems for years.

- Multi-core processors enable the same benefit as multiprocessors except at a reduced cost.

I remembered my graduate school days in the lab when banks of machines were fully utilized for the graphics students' ray-tracing project. I replied back, "Well, many applications can benefit from the horsepower made available through multi-core processors. A simple example is image processing where the work can be split between the different cores."

The engineer then stated, "Yeah, I can see some applications that would benefit, but aren't there just a limited few?"

My thoughts went to swarms of typical computer users running word processors or browsing the internet and not in immediate need of multi-core processors let alone the fastest single core processors available. I then thought the following:

- Who was it that said 640 kilobytes of computer memory is all anyone would ever need?

- Systems with multiple central processing units (CPUs) have not been targeted to the mass market before so developers have not had time to really develop applications that can benefit.

I said, "This is a classic chicken-and-egg problem. Engineers tend to be creative in finding ways to use the extra horsepower given to them. Microprocessor vendors want customers to see value from multi-core because value equates to price. I'm sure there will be some iteration as developers learn and apply more, tools mature and make it easier, and over time a greater number of cores become available on a given system. We will all push the envelope and discover just which applications will be able to take advantage of multi-core processors and how much."

The engineer next commented, "You mentioned 'developers learn.' What would I need to learn – as if I'm not overloaded already?"

At this point, I certainly didn't want to discourage the engineer, but also wanted to be direct and honest so ran through in my mind the list of things to say:

- Parallel programming will become mainstream and require software engineers to be fluent in the design and development of multi-threaded programs.

- Parallel programming places more of the stability and performance burden on the software and the software engineer who must coordinate communication and control of the processor cores.

"Many of the benefits to be derived from multi-core processors require software changes. The developers making the changes need to understand potential problem areas when it comes to parallel programming."

"Like what?" the overworked engineer asked knowing full well that he would not like the answer.

"Things like data races, synchronization and the challenges involved with it, workload balance, etc. These are topics for another day," I suggested.

Having satisfied this line of questioning, my software engineering colleague looked at me and asked, "Well what about embedded? I can see where multi-core processing can help in server farms rendering movies or serving web queries, but how can embedded applications take advantage of multi-core?"

Whenever someone mentions embedded, my first wonder is – what does he or she mean by "embedded"? Here's why:

- Embedded has connotations of "dumb" devices needing only legacy technology performing simple functions not much more complicated than those performed by a pocket calculator.

- The two applications could be considered embedded. The machines doing the actual work may look like standard personal computers, but they are fixed in function.

I responded, "One definition of embedded is fixed function which describes the machines running the two applications you mention. Regardless, besides the data parallel applications you mention, there are other techniques to parallelize work common in embedded applications. Functional decomposition is one technique or you can partition cores in an asymmetric fashion."

"Huh?" the software engineer asked.

At this point, I realized that continuing the discussion would require detail and time that neither of us really wanted to spend at this point so I quickly brought up a different topic. "Let's not talk too much shop today. How are the kids?" I asked.

1.1 Motivation

The questions raised in the previous conversation include:

- What are multi-core processors and what benefits do they provide?

- What applications can benefit from multi-core processors and how do you derive the benefit?

- What are the challenges when applying multi-core processors? How do you overcome them?

- What is unique about the embedded market segments with regard to multi-core processors?

Many of the terms used in the conversation may not be familiar to the reader and this is intentional. The reader is encouraged to look up any unfamiliar term in the glossary or hold off until the terms are introduced and explained in detail in later portions of the book. The rest of this chapter looks at each of the key points mentioned in the conversation and provides a little more detail as well as setting the tone for the rest of the book. The following chapters expound on the questions and answers in even greater detail.

1.2 The Advent of Multi-core Processors

A *multi-core processor* consists of multiple *central processing units* (CPUs) residing in one physical package and interfaced to a motherboard. Multi-core processors have been introduced by semiconductor manufacturers across multiple market segments. The basic motivation is *performance* – using multi-core processors can result in faster execution time, increased throughput, and lower power usage for embedded *applications*. The expectation is that the ratio of multi-core processors sold to single core processors sold will trend even higher over time as the technical needs and economics make sense in increasing numbers of market segments. For example, in late 2006 a barrier was crossed when Intel® began selling more multi-core processors than single core processors in the desktop and server market segments. Single core processors still have a place where absolute cost is prioritized over performance, but again the expectation is that the continuing march of technology will enable multi-core processors to meet the needs of currently out-of-reach market segments.

1.3 Multiprocessor Systems Are Not New

A *multiprocessor system* consists of multiple processors residing within one system. The processors that make up a multiprocessor system may be single core or multi-core processors. Figure 1.1 shows three different system layouts, a single core/single processor system, a multiprocessor system, and a multiprocessor/multi-core system.

Multiprocessor systems, which are systems containing multiple processors, have been available for many years. For example, pick up just about any book on the history of computers and you can read about the early Cray [1] machines or the Illiac IV [2]. The first widely available multiprocessor systems employing *x86 processors* were the Intel iPSC systems of the late 1980s, which configured a set of Intel® i386™ processors in a cube formation. The challenge in programming these systems was how to efficiently split the work between multiple processors each with its own memory. The same challenge exists in

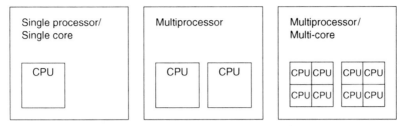

Figure 1.1: Three system configurations

today's multi-core systems configured in an asymmetric layout where each processor has a different view of the system. The first widely available dual processor *IA-32 architecture* system where memory is shared was based upon the Pentium® processor launched in 1994. One of the main challenges in programming these systems was the coordination of access to shared data by the multiple processors. The same challenge exists in today's multi-core processor systems when running under a *shared memory* environment.

Increased performance was the motivation for developing multiprocessor systems in the past and the same reason multi-core systems are being developed today. The same relative benefits of past multiprocessor systems are seen in today's multi-core systems. These benefits are summarized as:

- Faster execution time

- Increased throughput

In the early 1990s, a group of thirty 60 *Megahertz* (MHz) Pentium processors with each processor computing approximately 5 million floating-point operations a second (MFLOPS) amounted in total to about 150 MFLOPS of processing power. The processing power of this pool of machines could be tied together using an *Application Programming Interface* (API) such as Parallel Virtual Machine [3] (PVM) to complete complicated ray-tracing algorithms.

Today, a single Intel® Core™ 2 Quad processor delivers on the order of 30,000 MFLOPS and a single Intel® Core™ 2 Duo processor delivers on the order of 15,000 MFLOPS. These machines are tied together using PVM or Message Passing Interface [4] (MPI) and complete the same ray-tracing algorithms working on larger problem sizes and finishing them in faster times than single core/single processor systems.

The Dual-Core Intel® Xeon® Processor 5100 series is an example of a multi-core/multi-processor that features two dual-core Core™ processors in one system. Figure 1.2 is a sample embedded platform that employs this particular dual-core dual processor.

1.4 Applications Will Need to be Multi-threaded

Paul Otellini, CEO of Intel Corporation, stated the following at the Fall 2003 Intel Developer Forum:

> We will go from putting Hyper-threading Technology in our products to bringing dual-core capability in our mainstream client microprocessors over time. For the software developers out there, you need to assume that threading is pervasive.

This forward-looking statement serves as encouragement and a warning that to take maximum advantage of the performance benefits of future processors you will need to take action. There are three options to choose from when considering what to do with multi-core processors:

1. Do nothing
2. Multi-task or Partition
3. Multi-thread

Figure 1.2: Intel NetStructure® MPCBL0050 single board computer

The first option, "Do nothing," maintains the same legacy software with no changes to accommodate multi-core processors. This option will result in minimal performance increases because the code will not take advantage of the multiple cores and only take advantage of the incremental increases in performance offered through successive generations of improvements to the *microarchitecture* and the software tools that optimize for them.

The second option is to multi-task or partition. Multi-tasking is the ability to run multiple processes at the same time. *Partitioning* is the activity of assigning cores to run specific operating systems (OSes). Multi-tasking and partitioning reap performance benefits from multi-core processors. For embedded applications, partitioning is a key technique that can lead to substantial improvements in performance or reductions in cost.

The final option is to multi-thread your application. Multi-threading is one of the main routes to acquiring the performance benefits of multi-core processors. Multi-threading requires designing applications in such a way that the work can be completed by independent workers functioning in the same sandbox. In multi-threaded applications, the workers are the individual processor cores and the sandbox represents the application data and memory.

Figure 1.3 is a scenario showing two classes of software developers responding to the shift to multi-core processors and their obtained application performance over time. The *x*-axis represents time, and the *y*-axis represents application performance. The top line labeled "Platform Potential" represents the uppermost bound for performance of a given platform and is the ceiling for application performance. In general, it is impossible to perfectly optimize your code for a given processor and so the middle line represents the attained performance for developers who invest resources in optimizing. The bottom

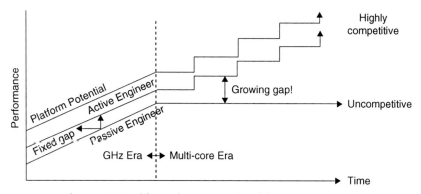

Figure 1.3: Taking advantage of multi-core processors

line represents the obtained performance for developers who do not invest in tuning their applications. In the period of time labeled the Gigahertz Era, developers could rely upon the increasing performance of processors to deliver end-application performance and the relative gap between those developers who made an effort to optimize and those that did not stayed pretty constant. The *Gigahertz* Era began in the year 2000 with the introduction of the first processors clocked greater than 1 GHz and ended in 2005 with the introduction of multi-core processors. Moving into the Multi-core Processor Era shows a new trend replacing that of the Gigahertz Era. Those developers who make an effort to optimize for multi-core processors will widen the performance gap over those developers who do not take action. On the flipside, if a competitor decides to take advantage of the benefits of multi-core and you do not, you may be at a growing performance disadvantage as successive generations of multi-core processors are introduced. James Reinders, a multi-core evangelist at Intel summarizes the situation as "Think Parallel or Perish." [5] In the past, parallel programming was relegated to a small subset of software engineers working in fields such as weather modeling, particle physics, and graphics. The advent of multi-core processors is pushing the need to "think parallel" to the embedded market segments.

1.5 Software Burden or Opportunity

On whose shoulders does the task of multi-threading and partitioning lie? It would be fantastic if hardware or software was available that automatically took advantage of multi-core processors for the majority of developers, but this is simply not the case. This instead is accomplished by the software engineer in the activities of multi-threading and partitioning. For example, in the case of multi-threading developers will need to follow a development process that includes steps such as determining where to multi-thread, how to multi-thread, debugging the code, and performance tuning it. Multi-threading places additional demands on the skill set of the software engineer and can be considered an additional burden over what was demanded in the past. At the same time, this burden can be considered an opportunity. Software engineers will make even greater contributions to the end performance of the application. To be ready for this opportunity, software engineers need to be educated in the science of parallel programming.

What are some of the challenges in parallel programming? An analogy to parallel programming exists in many corporations today and serves as an illustration to these challenges. Consider an entrepreneur who starts a corporation consisting of one employee, himself. The organization structure of the company is pretty simple. Communication and execution is pretty efficient. The amount of work that can be accomplished is limited due to the number of employees. Time passes. The entrepreneur has some success with venture capitalists and obtains funding. He hires a staff of eight workers. Each of these workers is assigned different functional areas and coordinate their work so simple problems such as paying the same bill twice are avoided. Even though there are multiple workers coordinating their activities, the team is pretty efficient in carrying out its responsibilities. Now suppose the company went public and was able to finance the hiring of hundreds of employees. Another division of labor occurs and a big multilayer organization is formed. Now we start to see classic organizational issues emerge such as slow dispersal of information and duplicated efforts. The corporation may be able to get more net work accomplished than the smaller versions; however, there is increasing inefficiency creeping into the organization. In a nutshell, this is very similar to what can occur when programming a large number of cores except instead of organizational terms such as overload and underuse, we have the parallel programming issue termed workload balance. Instead of accidentally paying the same bill twice, the parallel programming version is termed data race.

Figure 1.4 illustrates the advantages and disadvantages of a larger workforce, which also parallels the same advantages and disadvantages of parallel processing.

Advantages:

- Accomplish more by division of labor
- Work efficiently by specializing

©iStockphoto

Disadvantages:

- Requires planning to divide work effectively
- Requires efficient communication

Figure 1.4: Advantages and disadvantages of multiple workers

Many new challenges present themselves as a result of the advent of multi-core processors and these can be summarized as:

- Efficient division of labor between the cores

- Synchronization of access to shared items

- Effective use of the memory hierarchy

These challenges and solutions to them will be discussed in later chapters.

1.6 What is Embedded?

The term embedded has many possible connotations and definitions. Some may think an embedded system implies a low-power and low-performing system, such as a simple calculator. Others may claim that all systems outside of personal computers are embedded systems. Before attempting to answer the question "What is embedded?" expectations must be set – there is no all-encompassing answer. For every proposed definition, there is a counter example. Having stated this fact, there are a number of device characteristics that can tell you if the device you are dealing with is in fact an embedded device, namely:

- Fixed function

- Customized OS

- Customized form factor

- Cross platform development

A fixed function device is one that performs a fixed set of functions and is not easily expandable. For example, an MP3 player is a device designed to perform one function well, play music. This device may be capable of performing other functions; my MP3 player can display photos and play movie clips. However, the device is not user-expandable to perform even more functions such as playing games or browsing the internet. The features, functions, and applications made available when the device ships is basically all you get. A desktop system on the other hand is capable of performing all of these tasks and can be expanded through the installation of new hardware and software; it is therefore not considered fixed function.

The term, fixed function, may cause you to misperceive that embedded systems require only low-performance microcontrollers with certainly no need for multi-core processor performance. For example, consider a microcontroller controlling tasks such as fuel injection in an automobile or machines in a factory. The automotive and industrial segments are certainly two market segments that benefit from embedded processors. The reality is that the performance needs of these market segments are increasing as the fuel injection systems of tomorrow require the ability to monitor the engine with increasing response time and factory machines become more complicated. There is also the opportunity to consolidate functions that previously were performed on several microcontrollers or low-performance processors onto a fewer number of multi-core processors. In addition, the *automotive infotainment* market segment is impacting the automobile industry with the need to have several types of media and Internet applications inside a car. Thus, there is a demand for the performance benefits offered by multi-core processors.

Embedded devices typically employ OSes that are customized for the specific function. An MP3 player may be executing a version of Linux[*], but does not need capabilities that you would see on a desktop version of Linux, such as networking and X Windows capability. The OS could be stripped down and made smaller as a result including only the key features necessary for the device. This customization allows more efficient use of the processor core resources and better *power utilization*. Figure 1.5 is a screenshot of the System Builder interface used in QNX Momentics[*] to define the software components included in an image of QNX Neutrino[*] RTOS. Possessing a customized OS does not encompass all embedded devices. For example, consider a development team that uses a farm of desktop-class systems dedicated to performing parallel compilations. One could argue that these machines are embedded in that the functionality of these machines is fixed to one task; the machines are used to compile code and nothing else.

For the ease of maintainability, standard desktop OSes such as Windows[*] or Linux may be installed on these machines.

Embedded devices are typically customized for the desired use. A system used as a high-performance router contains many of the components used in a general server, but most likely does not contain all and, in addition, contains specialized components. For example, a customized router product may include customized cooling to fit inside space-constrained *1U* systems, solid state hard drives for temporary storage and high

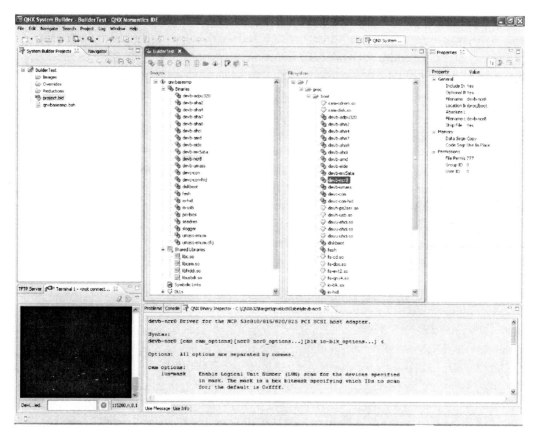

Figure 1.5: QNX Momentics system builder

availability, and specialized multiple Network Interface Cards (NICs) for performing its primary purpose.

One last feature of embedded systems concerns the method used by developers to program the system. A developer for general purpose computers can typically develop on the same system as those sold to customers. In other words, a program that executes under Windows XP on a laptop system may very well have been programmed on a laptop that runs Windows XP. This is typically not the case in embedded systems because the system is customized both in terms of OS and form factor so may not include the programming facilities necessary for development. For example, the likely development system for the media player included on the MP3 player is a GUI development environment running on a workstation class system. It may be theoretically possible to create a development environment that is hosted on the embedded system, but it would be very inefficient to ask the programmer to develop on a small 8 by 10 character LCD screen.

1.7 What is Unique About Embedded?

The utilization of multi-core processors in the embedded market segments offer unique challenges that are different than the challenges faced in the desktop and server market segments. These challenges are summarized as:

- Not all embedded OSes and the software tools available on these OSes fully support multi-core processors.

- Many legacy embedded applications do not move easily to modern multi-core processors.

- Embedded designs featuring converged and heterogeneous cores increase the programming and communication complexity.

- The introduction of Mobile Internet Devices based upon low-power x86 architectures provides new targets to support.

The embedded market segments employ a variety of different OSes that are very different than the desktop and server market segments where the majority of users employ a Microsoft Windows or a Linux OS.

Many embedded applications were developed on older 8-bit and 16-bit microprocessors and migrated to newer 32-bit microprocessors. Some multi-core processors offer 64-bit addressing. Older legacy embedded applications may have been coded with byte ordering assumptions making it difficult to move between architectures with *big endian* format and *little endian* format.

Embedded systems are typically specialized for the particular domain. The availability of transistors has led computer architects to design both increasingly complicated and increasingly specialized designs. One such complexity is the use of heterogeneous designs in the wireless market segment. A heterogeneous system contains different types of processors. For example, the Intel® PXA800F Cellular processor contains an Xscale™ processor and a Micro Signal Architecture digital signal processor inside the same package. Many of today's x86-based multi-core processors feature copies of the same class of CPU. In the future, heterogeneous x86-embedded designs will be introduced.

Another trend is the movement of x86 into new market segments such as Mobile Internet Devices. New low-power x86 processors are being introduced with embedded systems based upon them. This presents application developers targeting both multi-core processors and low-power x86 processors with the challenge of supporting both single core and multi-core targets. Single core processors may be disappearing in the desktop and server market segments, but the embedded market segments will continue to have them.

Chapter Summary

This chapter serves as an introduction to embedded multi-core processors. Chapter 2 studies basic system and microprocessor architecture and provides a history of Embedded Intel Architecture processors and a practical guide to understanding x86 assembly language, a skill required by any performance-sensitive x86-embedded software developer. In Chapter 3 multi-core processors are detailed with definitions of common terms, how to quantify performance of multi-core processors and further explanation of challenges associated with multi-core processors in the embedded market segments. Chapter 4 discusses issues in migrating to embedded multi-core x86 processors specifically focusing on tools available to assist with the unique challenges of multi-core processors. Chapter 5 details usability techniques and single processor optimization techniques that are prerequisite to any multi-core specific optimization. In Chapter 6, a process for developing multi-threaded applications is detailed. Chapter 7 discusses a case study on an image rendering application. Chapter 8 contains a case study where multi-threading is applied to an embedded networking application. In Chapter 9, virtualization and partitioning techniques are detailed along with specific challenges and solutions. Chapter 10 focuses on Mobile Internet Devices and contains a case study on power utilization. Chapter 11 summarizes and concludes the book.

References

[1] Cray Incorporated, http://www.cray.com
[2] W. J. Bouknight, S. A. Denenberg, D. E. McIntre, J. M. Randall, A. H. Sameh, and D. L. Slotnick, The Illiac IV system. *Proc. IEEE*, *60*(4), 369–388, 1972.
[3] Parallel Virtual Machine, http://www.csm.ornl.gov/pvm
[4] Message Passing Interface, http://www.mpi-forum.org/
[5] J. Reinders, *Think Parallel or Perish*, http://www.devx.com/go-parallel/Article/32784

## CHAPTER 2

Basic System and Processor Architecture

Key Points

- Performance is the key set of characteristics used to compare competing systems. For this book, performance implies start-to-finish execution time unless otherwise stated.

- Embedded Intel® Architecture processors have evolved over 30 years offering increased performance through microarchitecture features and increased functionality through integration.

- Understanding assembly language is critical to performance tuning and debugging. A number of simple tips will help you analyze IA-32 architecture assembly language code.

Open the case of an embedded system and one of the more prominent items you will typically see is a motherboard to which various integrated circuits are attached. Figure 2.1 depicts the high level components that make up an embedded system. These integrated circuits are generally comprised of one or more processors, a chipset, memory, and Input/Output (I/O) interfaces. Embedded software developers focused on performance analysis and tuning require a fair amount of knowledge of the underlying hardware. Developers employing *commercial off-the-shelf* (COTS) hardware and a commercial operating system or a full-featured open source operating system require some understanding of the components that comprise an embedded system. This chapter provides system basics, a history on Embedded Intel® Architecture processors, and highlights key performance enhancing innovations introduced in the successive generations of x86 processors. The chapter concludes with a tutorial on reading and understanding x86 assembly

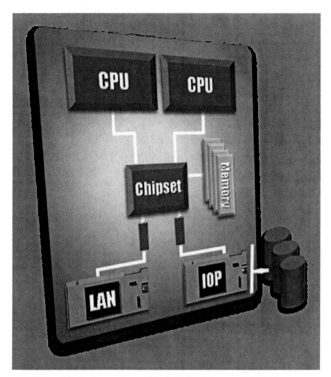

Figure 2.1: Sample embedded system components

language, a skill that all embedded developers focused on performance should know. Software developers who work on embedded systems with no operating system or a proprietary operating system require a very deep understanding of these components and are referred to the related reading section.

Figure 2.2 provides a more detailed schematic of a system layout.

The processors are the control unit of the system. Simply put, they are the "brain" of the system taking input from various sources, executing a program written in its instruction set that tells it how to process input, and subsequently sending the output to the appropriate devices.

The chipset interfaces the different components of the system together. Simply put, it is the "nervous system." It is the communication hub where the processor sends and

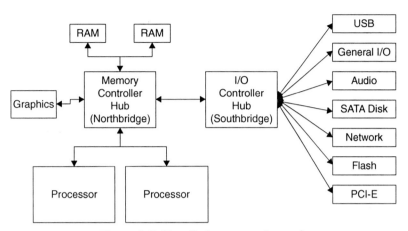

Figure 2.2: Detailed system schematic

receives information from the other components such as the memory and I/O on the system. In Figure 2.2, the chipset is represented by the memory controller hub (MCH), commonly termed "Northbridge" and the Input/Output controller (ICH), commonly termed "Southbridge."

The MCH has high speed connections to banks of memory. The memory is where programs and application data are stored while a particular application is active. It is the working area where the processor stores results during program execution. There are different types of memory used in embedded systems depending on the particular requirements. One optimization to increase the *bandwidth* of memory is to offer two paths to memory, termed dual channel. The MCH also typically controls communications to the graphics device.

The ICH controls access to relatively slower devices such as I/O and LAN interfaces. These interfaces control devices that provide input and output to the embedded system. Examples of I/O devices include Universal Serial Bus (USB) devices, keyboards, mice, touch screens, storage devices (such as Serial ATA drives), and the network.

2.1 Performance

The performance of an embedded system can have a different meaning depending upon the particular application of the device. In general terms, performance is whatever characteristic other similar devices compete upon and customers consider a key comparison point. Examples of performance in embedded systems include:

- Response time

- Start-to-finish execution time

- Number of tasks completed per unit of time

- Power utilization

In addition, many embedded systems have multiple performance needs. For example, an embedded system may have response time constraints as well as execution time constraints.

With regards to embedded systems and the impact that an embedded software developer has upon performance, the focus is either upon decreasing execution time or improving *throughput*. Execution time is the *latency* of a computation task. Throughput is a measure of how much work is accomplished per time unit. When performance is mentioned in further sections assume execution time is being discussed. Other performance measures will be mentioned explicitly.

One of the critical contributors to performance is the underlying processor of the embedded system. The history of Embedded Intel® Architecture processors is one of improving performance from generation to generation. The next section provides a timeline of several IA-32 architecture processors and discusses significant features that were introduced with each processor.

2.2 Brief History of Embedded Intel® Architecture Processors

One interesting fact is that the world's first processor was an embedded device. The Intel 4004 processor was developed in 1971 and targeted application in a business calculator. This device featured 2250 transistors and was a 4-bit processor capable of addressing 4096 bytes of memory.

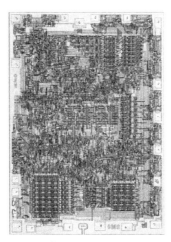

Intel 4004 Processor

2.2.1 Intel® 186 Processor

The Intel® 186 processor was introduced in 1982 with a clock speed of 6 MHz. The processor is similar to an Intel® 8086 processor, which is the processor used in the original IBM PC. The Intel 186 processor integrates a *Direct Memory Access* (DMA) controller and interrupt controller. *Integration* is the addition of functionality onto an embedded processor that was provided in separate integrated circuits in previous processors. The enhanced Intel 186 processor used today in embedded applications runs at 25 MHz and can address up to 1 MB of memory. The processor is found in many satellite control systems.

2.2.2 Intel386™ Processor

The Intel386™ processor was introduced in 1985 with a clock speed of 16 MHz and built with 275,000 transistors. This processor introduced a number of capabilities to x86 processors including 32-bit processing, protected memory, and *task switching*. Embedded versions currently in use range from 16 MHz to 40 MHz in clock speed and are capable of addressing between 16 MB and 4 GB depending on the specific model. Embedded applications of the processor vary from satellite control systems to robotic control systems. The Intel386™ is touted as a performance upgrade path for customers employing the Intel® 186 processor.

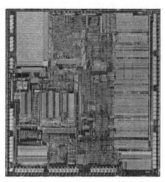

Intel386™ Processor

2.2.2.1 32-Bit Processor

A 32-bit processor is able to compute on data 32 bits at a time. Previous to the Intel386™ processor, x86 processors were 16 bit or 8 bit. 16-bit processors are relatively limited in the values that can be represented in its native format. A 16-bit value ranges from 0 to 65535. A 32-bit value can represent an integer between 0 and approximately 4 billion.[1] You can imagine some applications that would have difficulty if values were limited[2] to 65535; those same applications would function adequately with a range of 0 to 4 billion. The amount of memory addressable by the Intel386™ also meets the needs of the applications – how many applications exist that need to track over 4 billion items? For this reason, 32-bit processors have been sufficient for general purpose embedded computing for some time now.

A 32-bit processor can be extended to perform 64-bit arithmetic or can simulate 64-bit operations albeit at reduced speed compared to native 64-bit processors. Figure 2.3 shows a 64-bit addition performed on a 32-bit processor that occurs by breaking down

---

[1] 2^{32} is equal to 4294967296.

[2] Now it should be stated that 16-bit processors can compute on values larger than 16 bits by dividing the computations into 16-bit portions. This breaking down is similar to how we are taught to perform two digit multiplications by computing partial products and adding them together by ones, tens, hundreds, and so on.

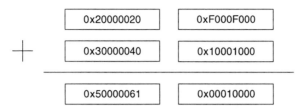

Figure 2.3: Emulation of 64-bit addition

31	16	8	0	79	0
EAX	AX/AH	AL		ST(0)	
EBX	BX/BH	BL		ST(1)	
ECX	CX/CH	CL		ST(2)	
EDX	DX/DH	DL		ST(3)	
ESP	SP			ST(4)	
EBP	BP			ST(5)	
ESI	SI			ST(6)	
EDI	DI			ST(7)	

Figure 2.4: x86 & x87 register set

the addition into a series of 32-bit additions. Software can emulate the addition paying special attention to the potential carry between the high end of the low 32 bits and the high 32 bits.

The basic IA-32 architecture register set contains *general purpose registers* (GPRs) entitled EAX, ECX, EDX, EBX, ESP, EBP, ESI, and EDI. Figure 2.4 depicts the registers in the basic IA-32 *instruction set architecture* (ISA). The 32-bit registers EAX, EBX, ECX, and EDX are accessible in 16-bit or 8-bit forms. For example, to access the low 16 bits of EAX in your assembly code, reference AX. To access the low 8 bits of EAX, reference AL. To access the second 8 bits of EAX or the high 8 bits of AX, reference AH. The Intel386 was often paired with an x87 *floating-point* coprocessor whose register set is also depicted. The x87 register set consists of eight 80-bit values accessed as a *stack*.

2.2.2.2 Protected Memory Model

The Intel386™ processor introduced the protected memory model that allows operating systems to provide a number of capabilities including:

- Memory protection

- Virtual memory

- Task switching

Previous to the Intel386™ processor, x86 operating systems were limited in terms of memory space and *memory protection*. Programs executing on a processor without memory protection were free to write to memory outside of its assigned area and potentially corrupt other programs and data. A program was essentially given control of the system and if it crashed, it brought the whole system down. Memory protection solves this issue by restricting the memory that a particular program can access to the region assigned to it by the operating system. If a program attempts to access memory outside of its assigned region, the processor catches the access and allows the operating system to intervene without allowing the memory to be changed, potentially corrupted, and bring the entire system down.

Virtual memory provides an application with an address space that appears to be contiguously labeled from 0 to 4 GB[3] even if physical memory is limited. The processor has functionality that maps addresses between the virtual address employed by the program to the underlying physical address on the system. Typically, a hard drive is used to simulate the larger address range in cases where the physical memory is less than the virtual address space.

Task switching or multi-tasking enables a modern operating system to execute multiple processes concurrently and switch between them based upon OS scheduling heuristics. The key technologies that enable task switching are the ability to interrupt a running process and functionality in the operating system to save the *runtime context* of the currently executing process.

---

[3] 4 GB in the case of 32-bit processors.

2.2.3 Intel486^{TM} Processor

The Intel486^{TM} processor was introduced in 1989 at a clock speed of 20 MHz and built with 1.2 million transistors. It is functionally similar to an Intel386^{TM} processor but integrates a floating-point unit and *cache* on the processor. In addition, this processor employs *pipelining* to increase performance. The processor has enjoyed a long lifespan and is still used in embedded applications such as avionic flight systems, *point-of-sales* (POS) devices, and *global positioning systems* (GPS). Current embedded versions of the Intel486^{TM} processor range in speed from 33 to 100 MHz. The Intel486^{TM} processor is a natural upgrade path to embedded device manufacturers employing the Intel386^{TM} processor.

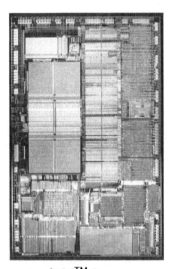

Intel486^{TM} Processor

2.2.3.1 Floating Point

The Intel486^{TM} processor integrates a floating-point unit. This floating-point unit was equivalent to the 387 coprocessor that is available as a supplement to the Intel386^{TM} processor. A floating-point unit performs operations on floating-point numbers. Floating-point numbers are used to represent real numbers (positive and negative numbers having a decimal point and numbers following the decimal point). An example floating-point number is 3.14156. The x87 floating-point model is stack based with 8 registers that are each 80 bits in size (see Figure 2.4). Floating-point values are loaded from memory where they are stored as either 32 bit (single precision) or 64 bit (double precision). Typically one of the operands will be the top of the stack and is popped, operated on, and then pushed back onto the stack. Integrated floating-point support enables high performance execution of many different embedded applications, such as instrumentation, aviation, and navigation.

2.2.3.2 Cache Memory

Cache memory is a high speed memory that sits between the processor and main memory and provides fast access to a reservoir of code and data. Cache memory is necessary because the speed attainable by processors has increased relative to the speed of main memory. Without cache, the processor would sit idle for many cycles waiting for data to be read from and written to main memory. The original Intel486™ processor features an integrated 8K cache that provides both data and instructions. All current x86 processors feature several levels of cache memory containing several *Megabytes* (MB) of memory. Table 2.1 provides typical access times and sizes for the different elements of what is termed the *memory hierarchy*. As you can see, access times to memory that is farther away from the processor increases substantially. Effectively using cache is one of the best techniques for increasing application performance and is a primary focus in many performance analysis studies.

2.2.3.3 Pipelining

Pipelining is a technique used to increase processor performance and is motivated by the real world example of an assembly line. In an assembly line, the task to be completed is

Table 2.1: Typical memory hierarchy access time

Memory class	Access time	Typical size
Register	1 cycle	<1 KB
Level 1 cache	1–4 cycles	8–64 KB
Level 2 cache	5–30 cycles	512 KB–4 MB
Memory	Hundreds of cycles	512 MB–32 GB
Disk	Millions of cycles	40–500 GB

broken into a number of stages accomplished by different workers. One advantage of the assembly line is that multiple tasks can be in progress at different steps of the assembly line at the same time. The same holds true in a processor pipeline.

The Intel486™ processor pipeline breaks the execution of one x86 integer instruction into the following suboperations:

- Fetch

- Decode-1

- Decode-2

- Execute

- Write Back

The fetch stage obtains an instruction from memory. The decode-1 stage determines the operation to be performed and the required resources. The decode-2 stage determines the addresses of the operands for the operation. The execute stage is where the actual operation is performed on the values. The write back stage is where results are placed into their destination. With this breakdown of processor execution into stages, multiple instructions can be in progress at a given time. Figure 2.5 illustrates how five instructions execute through the various stages. The columns denote the time in cycles and are labeled one through eight. The rows are labeled "Instruction1" through "Instruction4" and represent sample instructions. The entries in the table represent the stage in the pipeline that the instruction is currently in with "F," "D1," "D2," "E," and "wb" corresponding

Time	1	2	3	4	5	6	7	8
Instruction 1	F	D1	D2	E	wb			
Instruction 2		F	D1	D2	E	wb		
Instruction 3			F	D1	D2	E	wb	
Instruction 4				F	D1	D2	E	wb

Figure 2.5: Pipelining of instructions

to Fetch, Decode-1, Decode-2, Execute, and Write Back. The advantage of pipelining is observed at time 5 through 8. Once the pipeline is loaded with instructions, on every cycle one instruction is completed. Consider a processor that is not pipelined that performs the same basic actions in decoding, executing, and writing back an instruction. In this case, one instruction is retired every 5 cycles. Conceptually, the benefit of pipelining is quite clear; however, there are many caveats to this simple example that prevents a true pipeline from achieving its theoretical maximum. Some of these caveats are dependencies between instructions, memory latency, and lack of execution resources, to name a few. Assembly language programmers and compiler writers need to be very aware of the pipeline when it comes to scheduling instructions. Embedded software developers need some awareness of pipelining, but optimal scheduling is more the domain of an aggressive optimizing compiler.

2.2.4 Intel® Pentium Processor

The Intel® Pentium® processor was introduced in 1993 and manufactured on a 0.8 micron process using approximately 3 million transistors. The microarchitecture was a new design that incorporated *superscalar execution* to increase performance. A subsequent version, the Intel Pentium processor with MMX™ technology is still used today in embedded applications such as POS systems, kiosks, and networking. Current versions of the processor operate at clock speeds ranging from 166 to 266 MHz and *Thermal Design Power* (TDP) of 4.1 to 17.0 W. *MMX™ instructions* are an extension to the IA-32 ISA focused on increasing the performance of integer processing. The Pentium processor was also the first Embedded Intel Architecture processor to feature *performance monitoring counters* (PMCs).

Pentium® Processor

2.2.4.1 Superscalar Execution

Superscalar execution is a performance feature of processors that allow multiple instructions to execute at the same time by duplicating execution resources. In the case of a processor that employs pipelining, a superscalar version of it would have multiple pipelines capable of executing at least two different instructions passing through the same stages of the pipeline at the same time.

Figure 2.6 shows a sample set of instructions executing on a pipelined superscalar processor. Superscalar execution requires the duplication of execution resources inside a processor. The advantage of superscalar execution is evident by comparing Figures 2.5 and 2.6. Instead of executing one instruction per cycle, the superscalar example is capable of executing two instructions per cycle.

The Pentium processor featured what are termed a U pipeline and a V pipeline. The V pipeline could execute certain instructions at the same time as the U pipeline under certain constraints. These constraints required the assembly programmer or the compiler to pay particular attention to the pairing of instructions so that they could execute together. Explicit scheduling of instructions is less of a concern in processors that feature *out-of-order* (OOO) *execution*, which is discussed in a later section.

2.2.4.2 Performance Monitoring Counters

PMCs collect information on internal processor events and enable software tools to record and display the collected information. This information consists of counts of events like clock cycles, instructions retired, and *cache misses*. The Pentium processor

Time	1	2	3	4	5	6	7	8
Instruction 1	F	D1	D2	E	wb			
Instruction 2	F	D1	D2	E	wb			
Instruction 3		F	D1	D2	E	wb		
Instruction 4		F	D1	D2	E	wb		

Figure 2.6: Superscalar execution pipeline

was the first Embedded Intel Architecture processor to enable basic counting of events. The technology has increased in capability and number of processor events that can be recorded in successive processor generations. For example, it is now possible to correlate a given event such as cache miss with the instruction that caused the miss. Embedded software developers focused on performance tuning should employ PMC measurement tools to help tune their applications. Section 4.3 details tools that take advantage of PMC.

2.2.5 The Intel® Pentium III Processor

The Intel® Pentium III processor was introduced in 1999 and is based upon previous generation Pentium II and Pentium Pro processors. At introduction, the clock speed was 500 MHz built using a 0.18 micron process and utilizing 9.5 million transistors. The processor integrates even larger amounts of cache memory (512K in some versions), incorporates OOO execution, and introduces *Streaming SIMD Extensions* (SSE), an instruction set extension geared toward high performance floating-point computation. Current versions of the Pentium III used in embedded devices range from 600 MHz to 1.26 GHz. Current embedded applications include industrial applications such as data acquisition, kiosk, gaming, and security.

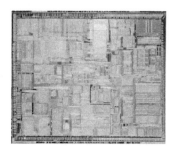

Pentium® III Processor

2.2.5.1 OOO Execution

OOO execution is a feature of processors where instructions are allowed to execute in an order different than the sequential order that is specified in the *machine language* representation of the program. The processor contains logic that allows instructions with no outstanding dependencies on other in-progress instructions to execute. Figure 2.7 shows assembly language for two additions and the dependencies between instructions. Specifically, even though the instructions at line 4 and 5 are fourth and fifth in the listing,

```
Line  Assembly Code
 1    movl      $1, %eax
 2    movl      $1, %ecx
 3    addl      %eax, %ecx
 4    movl      $1, %ebx
 5    movl      $1, %edx
 6    addl      %ebx, edx
```

Figure 2.7: Sample assembly language program

a superscalar processor could consider these instructions for execution at the same time as the instructions at line 1 and 2 assuming there are enough execution resources to accommodate. Notice the instruction at line 3 is dependent upon the instructions at line 1 and 2. Similarly, the instruction at line 6 is dependent upon the instructions at line 4 and 5. A superscalar processor respects the dependency and so would not execute the instructions at line 3 and line 6 until the data values were ready.

2.2.5.2 Streaming SIMD Extensions

SSE is an extension to the existing x86 and x87 instruction sets that enables *Single Instruction Multiple Data* (SIMD) execution of single precision floating-point values. SIMD processing is also known as vector processing and allows the same computation to be performed on multiple pieces of data (Multiple Data) using only one instruction (Single Instruction). The SSE instruction set features 70 instructions operating on single precision (32 bit) floating-point values common in the area of image processing, media encoding, and media decoding. SIMD execution can increase performance tremendously for these types of applications as they tend to employ common operations on vast amounts of data. Subsequent extensions of the instruction set have been introduced including SSE2, SSE3, SSSE3, and SSE4. SSE added 8 new 128-bit registers called xmm0–xmm7 capable of holding 4 single precision floating-point values or 2 double precision floating-point values. Employ these instructions to substantially increase the performance of your application code. Table 2.2 summarizes the evolution of SSE since introduction.

2.2.6 The Intel Pentium® 4 Processor

The Intel® Pentium 4 processor was introduced in May 2000 and employs what is termed Intel NetBurst® microarchitecture which was optimized for high clock speeds and media applications. At introduction, the clock speed was 1.5 GHz using a 0.18 micron process and approximately 42 million transistors. The processor employs *hyper-threading*

Table 2.2: Instruction set extension description

	SSE	SSE2	SSE3	SSSE3	SSE4
Introduction	1999	2000	2004	2006	2007
# Instructions	70	144	13	32	47
Description	Single-precision, streaming operations	Double-precision, 128-bit vector integer	Complex arithmetic	Audio & video decode	Video accelerators, graphics building blocks

technology, a term for its *simultaneous multi-threading* (SMT), to increase performance. Current embedded versions of the processor range from a 1.7 GHz and 35 W TDP version to a 3.4 GHz and 84 W TDP. Current embedded applications of the Pentium 4 processor include sophisticated interactive clients, industrial automation solutions, and digital security surveillance (DSS). These applications though are migrating to more recent processors because of improved performance and power utilization.

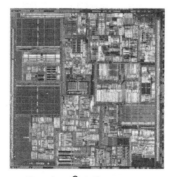

Pentium® 4 Processor

2.2.6.1 Hyper-threading Technology

Hyper-threading technology is the term used to describe Intel's version of SMT. SMT allows a single processor to appear as multiple processors to the operating system. The motivation for this technique came from the observation that even OOO execution is unable to use all of the processing power in the superscalar pipelines; there are limits to the amount of *parallelism* existing in one execution context. SMT attempts to use those resources by finding work in other *threads* and processes. Figure 2.8 shows how the

Hyper-threading Technology Resource Utilization

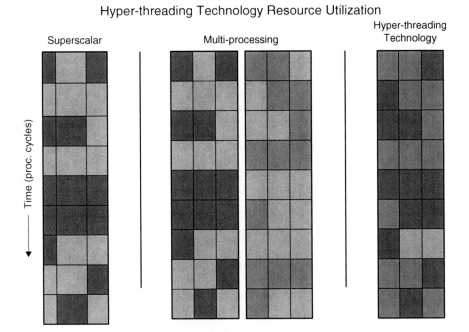

Note: Each box represents a processor execution unit

Figure 2.8: Hyper-threading execution model

resource utilization of hyper-threading technology differs from superscalar execution and multiprocessing. With superscalar execution, only the instructions associated with one process are in the execution pipeline at a time. With multiprocessing, instructions associated with two processes are executing, but each on a separate processor. Hyper-threading technology enables instructions associated with two processes to share the execution resources of one processor at the same time. Embedded software developers should pay attention to the sharing of data between threads. Section 6.2.4 provides more information on optimizing memory usage.

2.2.7 The Intel Pentium® M Processor

The Intel® Pentium M processor was introduced in 2002 and was based upon the best pieces of the previous generation Pentium III processor and Pentium 4 processor with a goal of decreasing power utilization. The clock speed at introduction ranged from 900 MHz to 1.7 GHz depending on the version. Current embedded versions of the processor range from 1.1 GHz and 12 W TDP to 2.0 GHz and 27 W TDP. Current embedded applications of the processor include sophisticated interactive clients, industrial automation, and enterprise communications.

Processor: Pentium® M Processor

2.2.7.1 Power Utilization

Comparing the TDP of the Pentium 4 processor and the Pentium M processor shows a fairly considerable decrease. In fact, the Pentium M processor was designed specifically for the mobile market and features were added to decrease overall power utilization such as C-states, micro-op fusion, and extended stack pointer folding. Enhanced Intel SpeedStep® technology [1] employs C-states, which are low power modes of a processor where the voltage and frequency are scaled down in cases of power critical applications that do not require the full processing capability. Micro-op fusion is the binding of two instructions together for execution in the pipeline that reduces the overhead. Extended stack pointer folding reduces the load on the execution units by handling stack adjustments separately. Power utilization has always been a concern in small embedded devices and is becoming an increasing constraint in high performance designs. Processor technology is approaching limits in scaling the frequency of processors while maintaining reasonable power utilization; this limit is referred to as the "Power Wall."

2.2.8 Dual-Core Intel Xeon® Processors LV and ULV and Dual-Core Intel® Xeon® Processor 5100 Series

The Dual-Core Intel Xeon® processors LV and ULV & Dual-Core Intel® Xeon® processor 5100 series were introduced in 2006 and are based upon the desktop/server processor known as Intel Core processor. The processors are clocked from 1.66 to 2.33 GHz, feature a 2 MB shared cache, and are a dual processor and dual-core design. Current embedded applications of the processors include storage area networks (SAN), network attached storage (NAS), and virtual private networks (VPN).

2.2.9 Intel® Core™ 2 Duo Processors for Embedded Computing

The Intel® Core™ 2 Duo processors for Embedded Computing were introduced in 2006 and were a refinement of the previous generation Dual-Core Intel® Xeon® processors LV and ULV. The clock speed of the processor ranges from 1.5 to 2.16 GHz, features 2 or 4 MB of shared cache, and is a dual-core design. TDP for the processor range from 17 W for the 1.5 GHz version to 64 W for the 2.13 GHz version with a front-side bus speed of 1066 MHz. Current embedded applications of the processor target smaller form factors than the previous processor such as interactive clients, gaming platforms, DSS, and medical imaging. This processor also introduced Intel® 64 ISA support, which is a 64-bit instruction set and provides a larger address range than previous 32-bit processors.

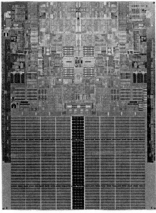

Intel® Core™ 2 Duo Processors for Embedded Computing

2.2.9.1 Intel® 64 ISA

When Intel extended the instruction set of the IA-32 ISA to support 64-bit computing, a number of changes occurred as a result: the first change is the increase in the internal size of its GPRs to 64 bit. The number of GPRs also increased from the 8 registers in the 32-bit x86 model to 16 registers with designations R8 to R15 used for the high 8 registers. In addition, the number of SSE registers was increased from 8 to 16. The address range was also extended to support 64-bit addresses. To ease migration to Intel 64 architecture, different execution modes were created. The first mode is called legacy mode and the second is called IA-32e mode. In legacy mode, 64-bit support is turned off and the processor is effectively a 32-bit processor. IA-32e mode supports the execution of both pure 64-bit applications and 32-bit applications. A few other noteworthy features include:

- Relative instruction pointer (RIP) addressing – access to data can be relative to the instruction pointer which eases development of position-independent code (PIC) for shared libraries.

- Addressable access to 32-bit, 16-bit, 8-bit components of all general purpose 64-bit registers.

2.2.10 Quad-Core Intel® Xeon® Processor 5300 Series

The Quad-Core Intel® Xeon® processor 5300 series was introduced in 2007, features four processor cores on a package, and is part of the Core 2 processor family. The clock

Quad-Core Intel® Xeon® Processor 5300

speed of the processor ranges from 2.0 to 2.33 GHz and features 8 MB of L2 cache. TDP for the processors is 80 W with a front-side bus speed of 1333 MHz. Current embedded applications of the processor target high performance embedded applications such as Intrusion detection/prevention, Services over IP (SoIP), and Video on Demand.

2.3 Embedded Trends and Near Term Processor Impact

Future processor trends call for increased integration of functionality, an increase in the number of processor cores, and continued focus on power utilization. Typical IA-32 architecture and Intel 64 architecture processors require two other separate chips to provide I/O, main memory, and graphics memory interfaces. *System-on-a-chip* (SOC) IA-32 architecture processors will integrate these functions due to the availability of transistors and will result in better power utilization. New lower power IA-32 architecture processors will be developed that optimize power by removing high performance processor features such as OOO superscalar execution and returning to a simpler pipelined architecture. Another future processor will feature a heterogeneous design teaming an IA-32 architecture processor with a set of packet processors.

One other trend worth mentioning is the continued adoption of leading edge IA-32 architecture processors targeting the desktop and server market segments. Multi-core processors built using 45 nm process technology are available for use in desktop and server systems. At the time of this writing, processors built using 45 nm process technology and targeting the embedded market segments are not available, but it is expected to be available by the time this book becomes available.

2.3.1 Future 45 nm Embedded Processor

The Intel® Xeon® 5400 processor has been announced for the server market segments and this processor or a modified version of it will eventually find application in the embedded market segments. The processor is a quad core design featuring 820 million transistors built on 45 nm process technology with up to 12 MB of level 2 cache. Also introduced with this processor are new Intel® SSE4 instructions, new techniques to reduce power when idle, and a new dynamic acceleration technology that increases the clock speed of one core for improved performance when the other cores are idle.

2.3.2 Intel® Atom™ Processor Core

A new processor, under development at the time of this writing and called the Intel® Atom™ Processor, is designed for application as a low power processor suitable for fanless embedded systems and featuring a simpler pipeline and new features to reduce power utilization. Some details on the processor core include:

- Built using 45 nm process technology.

- Employs in-order execution instead of OOO execution which reduces potential performance, but saves power.

- Employs superscalar execution with the ability to retire two instructions per cycle.

- Employs Hyper-threading Technology.

- Includes Intel® 64 ISA support.

- Adds a new low power sleep state called C6.

The Intel® Atom™ Processor is one of the building blocks for future Mobile Internet Device (MID) platforms that offer a combination of processor core and chipset. For example, the Moorestown platform features a low power Intel Architecture processor core with integrated memory controller and video acceleration features. The platform features an enhanced I/O chip termed a communications hub for low power and accelerated I/O operations.

2.3.3 Tolapai SOC Accelerator

The Tolapai SOC Accelerator integrates the memory controller and I/O controller onto the processor, which has not been done for an Embedded Intel® Architecture processor since the Intel386™ EX processor. The IA-32 processor core on the integrated package is a derivative of an Intel® Pentium® M processor with a 256K L2 cache. Integration also includes support for Intel® QuickAssist Technology which employs a set of packet processing engines acting as *accelerators* for functions such as content processing and security in networking applications. Some applications can benefit by as much as 8 times the performance of an IA-32 processor core alone.

2.4 Tutorial on x86 Assembly Language

Assembly language is a representation of the machine language code that makes up an application. Reading and understanding x86 assembly language are difficult tasks. Writing x86 assembly language is even more difficult. Embedded software engineers are daunted by the sheer longevity of the instruction set and breadth that includes:

- 8-bit, 16-bit, 32-bit, and 64-bit general purpose register sets

- Variable length instructions

- Multiple instructions that perform the same action

- Aggressive compiler optimization

- New instructions and register sets

- x87 floating point

Regardless, there are times when an embedded software engineer will need to look at a disassembly of an application binary or the source code compiled into assembly to perform debugging or performance tuning. For these reasons a tutorial is provided that will equip the reader with the competence necessary to perform these actions. The tutorial first covers basics on x86 assembly language and then details seven tips that will help you gain familiarity. The section concludes with an analysis of a small C program compiled to assembly language.

2.4.1 X86 Assembly Basics

The first task in understanding assembly language is to know how to easily generate it. Two techniques for generating assembly for your application code are:

- Disassemble the application binary

- Compile the source code to assembly language.

Disassembly of an application binary requires a tool called a disassembler which maps the machine language contained in the binary to the equivalent assembly language mnemonics. A disassembler is a common tool on modern operating systems. On Linux systems, the *objdump -d* command produces a disassembly listing. For example, to disassemble the ls command, type:

objdump -d 'which ls

The disassemble command is not the same on all systems. On Windows and Mac OS, the disassemble command is *dumpbin /disasm* and *otool -t -V*, respectively. Many debuggers feature the option to view the code under debug as assembly language and is effectively calling a disassembler to display the instructions. To produce an assembly file with a compiler, consult your compiler's manual to find the correct option. Using gcc and other compatible compilers[4] on a Linux system, the option is -S. Compiling the code in Figure 2.9 with the Intel Compiler produces an assembly listing, a portion of which is in Figure 2.10.

The four components of an assembly listing are:

1. Instructions

2. Directives

3. Comments

4. Labels

Instructions map 1-to-1 to the low level machine language instructions that the processor executes. An example instruction is:

addl %eax, %ecx

```
int simple_loop(int x)
{
 int i, ret_val = 0;
 for (i=0; i< x; i++) {
   ret_val += i;
 }
 return ret_val;
}
```
Figure 2.9: Simple C function

---

[4] The Intel® C++ Compiler is compatible with gcc and accepts many of the same command line options.

```
..B1.3:
        addl      %eax, %ecx      # Preds ..B1.3 ..B1.2
        addl      $1, %eax        #5.3
        cmpl      %edx, %eax      #4.15
        jl        ..B1.3          #4.2
        .align    2,0x90          # Prob 82%   #4.2
```
Figure 2.10: Assembly language listing of simple function

which represents a 32-bit addition of register ecx and register eax. The result of the operation is stored in register ecx. The IA-32 ISA specifies explicit two operand format for operations on the GPRs. Two operand format means that for operations involving two source operands, one of the operands specifies the destination location. Three operand format is common on other architectures and allows explicit naming of two source locations and a destination that may be different. MMX, SSE, and its successive generations of instructions employ a two operand format. X87 floating point is stack-based and employs one explicitly named operand with the result being stored implicitly on top of the stack.

Directives specify commands to the assembler that control its generation of machine language code. The directive used in Figure 2.10 is the align directive that specifies the assembly to align the next instruction on a particular address boundary. The assembler has the ability to know how many bytes previous instructions have taken and can pad the address range with *nops* so a particular instruction is aligned. Alignment is a performance optimization for the instruction cache. It is generally good to align the target of a branch to guarantee it is the start of a cache line. Other directives include type, size, data, and section.

Labels are generic identifiers for locations in the program that can either be data or instructions. In Figure 2.10, the label denoted *..B1.3* is an identifier for the location of the instruction that immediately proceeds it. Labels identify portions of the code that are not explicitly known until the assembly code is assembled or until the machine language code is loaded for execution. For example a jump to a label requires either the absolute address or a relative offset of the target location. The absolute address will not be known until the operating system assigns addresses for the program. It is possible to compute the relative offset of a branch after assembly. For example, the branch in Figure 2.10 denoted *jl* would most likely use a relative offset because the number of bytes between it and the target address is the number of bytes comprised by the instructions in between, something that can be computed during assembly.

Comments are denoted by the # symbol and indicate extra information that is ignored by the assembler. In the case of the assembly listing in Figure 2.10, comments provide information on source line and computed branch probabilities. Source line information is particularly valuable because this value tells you the line number of the C++ source language statement that resulted in the assembly language instruction.

Only time and effort spent dissecting assembly code listings and writing assembly code will make you an expert on the subject. The following tips provide a minimal set of need-to-know essentials when reading x86 assembly language. Employ these tips when looking at the assembly listing of your programs.

2.4.2 Tip #1 – Focus on Small Regions

The first real tip is to avoid reading assembly language if you can get away with it. Reading and understanding assembly language is a time-consuming process so other lower hanging fruit for your optimization or debug effort should be employed before resorting to viewing assembly language. If you must study the assembly listing, focus your time on the most relevant and smallest portions possible. This is similar to performance tuning where you should focus on the frequently executed sections. Spend time on the critical few regions because the time investment can be quite high.

2.4.3 Tip #2 – Quickly Identify Source and Destination

There are different notations or formats for x86 assembly language. The formats are known as AT&T Unix notation and Intel notation. One of the key differences between them is the location of the source operand and destination operand. In AT&T Unix notation, the source comes before the destination reading left to right. In Intel notation, the destination comes before the source. A quick way to determine source operand location is to look for an operation involving a constant. In Figure 2.11 there are two

```
mov eax, 1    # Must be Intel notation
movl $1, %eax # Must be AT&T Unix notation
```
Figure 2.11: Sample assembly instructions[5]

---

[5] Two other differences are the syntax for referencing constants and for referencing registers.

different instructions, one labeled Intel syntax and the other AT&T Unix. It is impossible for a constant value to be the destination operand so you immediately know that the direction is left-to-right and right-to-left, respectively.

2.4.4 Tip #3 – Learn Basic Registers and Memory References

Understand the basic register set of the architecture and the different formats for referencing memory. Two challenges with x86 assembly language are (1) registers have different bit size versions and (2) there are multiple indirections possible when referencing memory. Learning the registers is just a matter of recognizing them when you see them. Recognizing memory references takes practice. A reference can include a base register, an index register, a scale, and an offset to compute the final location. Table 2.3 shows several example memory references and describes them.

2.4.5 Tip #4 – Learn Basic Frequently Used Operations

Much like learning a foreign language, if you understand the commonly used words, you'll be able to piece together portions of the meaning without understanding each and every word. The commonly used processor operations can be categorized as:

- Moving values
- Arithmetic
- Testing conditions
- Branches

Table 2.3: Sample memory and register references

Memory reference	Description
Object	Value stored at memory location object
$object	Address where object is stored
(%eax, %esi)	The contents of memory whose location is computed by adding the contents of register %esi to register %eax
24(%edx, %edi, 8)	The contents of memory whose location is computed by multiplying the contents of register %edi by 8, adding register %edx, and 24

The most common method of moving an integer value is the *mov* instruction which moves a value from the source location to the designation. In addition, there is a size specifier appended to the end of the instruction.[6] The "l" in "movl" specifies a move of a 32-bit value. A "w" that would be seen in a "movw" specifies a 16-bit move. Finally, an "s" that would be seen in a "movs" instruction specifies an 8-bit move.

The instruction to load an x87 floating-point value is simply *fld* and to move a value from one register location to another or to the top of the floating-point stack is *fswap*. For the media instructions, a number of instructions beginning with the string mov are used to move values to and from memory and registers. Table 2.4 lists several example move operations on different data types and provides an explanation.

Learn to identify basic common arithmetic operations. These are addition, subtraction, multiplication, and division on integers and floating-point values. Sample arithmetic instructions mnemonics are provided in Table 2.5.

Table 2.4: Sample move instructions

Command	Description
movl %eax, %edx	Moves the 32-bit value in register eax to register edx
flds v	Loads a 32-bit value on top of the x87 floating-point stack
movsd %xmm0, v	Moves the scalar double precision value of SSE register xmm0 to v
movups (%edx), %xmm3	Moves unaligned packed single-precision value whose address is located in register edx to register xmm3

Table 2.5: Sample arithmetic instructions

Operation	Integer	x87 Floating-Point	SSE (single precision)	SSE (double precision)
Addition	add	fadd	addps	addpd
Subtraction	sub	fsub	subps	subpd
Multiplication	mul	fmul	mulps	mulpd
Division	div	fdiv	divps	divpd

---

[6] We are assuming AT&T Unix syntax in all of the examples unless otherwise stated.

Learn the common test and compare operations and how to identify which bits are being tested. Sample test and compare instructions and explanations are provided in Table 2.6.

Learn branch instructions; sample branch instructions and explanations are provided in Table 2.7. The first two branch instructions are conditional and take the jump depending on the value of specific condition bits set after a previous compare operation or mathematical operation. The last two branch instructions are unconditional branches and are not dependent on condition bits.

2.4.6 Tip #5 – Your Friendly Neighborhood Reference Manual

The IA-32 architecture and Intel® 64 architecture reference manuals [2] provide an exhaustive list of every IA-32 architecture and Intel 64 instruction supported and details on each. Use this manual when you see an instruction in a critical region that you do not understand. The manual is also helpful for identifying instructions that set condition registers. For example the code in Figure 2.12 contains a branch instruction that branches if a certain condition bit is set. If you look up the semantics of the *cmpl* instruction that occurs previously in the listing, you will see that the compare instruction compares two

Table 2.6: Sample compare and test instructions

Command	Description
cmpltpd %xmm2, %xmm4	Determines if register xmm2 is less than xmm4 as a packed double and sets appropriate flags based upon result
fcomp %st(4)	Compares the top of the floating-point stack with stack position 4 and sets condition codes in FP status word based upon results
test $0, %eax	Performs a logical AND of 0 and register eax and set condition codes

Table 2.7: Sample branch instructions

Command	Description
jl	Jump if condition bits indicate less than (SF <> OF), short jump
je	Jump if condition bits indicated equality (the Zero Flag = 1), near jump
jmp	Direct jump – jump to address
jmp (%eax)	Indirect jump – jump to the address that is stored in register %eax

```
cmpl    %edx, %eax
jl      ..B1.3
```

Figure 2.12: Assembly language compare instruction

values and sets the condition bits SF and OF which are is used by the following branch instruction.

2.4.7 Tip #6 – Beware of Compiler Optimization

Aggressive compiler optimization can transform C or C++ code into very difficult to read assembly language for several reasons. The following common optimizations result in assembly code that is difficult to correlate to the C source code:

- Strength reduction

- Alternate entries

- Basic block placement

- Inlining

Strength reduction is the replacement of one type of operation to a different set of operations that are faster to execute. Some examples of strength reduction are the changing of multiplication to a series of shifts and additions, transforming exponentiation into multiplication, transforming division into a reciprocal multiplication, and initialization using less expensive operations. Figure 2.13 is a code example and resulting assembly language that shows a strength reduction where an integer multiplication by 4 is turned into a shift left of two bits. Also included is the setting of a value to 0 that is accomplished using the logical operation, *xor*.

Alternate entries occur as a result of an optimization that reduces the call and return overhead by allowing the calling routine to keep values in registers and jumping to a point of a called function where the registers have the expected values. The IA-32 ISA ABI specifies that function call arguments are to be placed on the stack. The called function would then move these items off of the stack into registers which essentially costs multiple moves to and from memory. Figure 2.14 shows the assembly language for a function that contains the normal function entry point and an alternate entry. Notice at the normal function entry point (label `multiply_d`) the first instructions move the values from the stack into registers whereas the alternate entry (label `multiply_d.`) assumes the values are already there.

```
int f, g;
f *= 4;
g = 0;


shll        $2, %eax
xor         %ebx, %ebx
```
Figure 2.13: Strength reduction example

```
     multiply_d:
..B4.1:               # 1              # Preds ..B4.0
     movl        4(%esp), %eax                        #11.1
     movl        8(%esp), %edx                        #11.1
     movl        12(%esp), %ecx                       #11.1

 multiply_d.:                                         #
     pushl       %edi                                 #11.1
```
Figure 2.14: Alternate entry code example

Basic block placement is an optimization for the instruction cache that attempts to place basic blocks that execute close together in time as close together address-wise as possible. A basic block is a set of instructions where if one instruction executes, then program flow dictates all of the other instructions will execute. Typically a basic block begins as the result of being a branch target and ends as a result of a branch instruction. If basic block placement is performed over a large function, basic blocks that make up a tiny portion of the function can be scattered throughout the assembly language listing for the function. The key for determining the logical flow of your program to identify such things as loops is to create a graph of basic blocks. Figure 2.15 is C source code for a switch statement where the comments indicate the frequency of execution for each case. Figure 2.16 shows how a compiler may rearrange the order of the assembly language for each of the case statements based upon measured execution frequency. As you can see, the compiler placed the assembly language code associated with the "case 10" code that occurs 75% of the time to be the first case tested. The compiler can determine execution frequency by employing profile-guided optimization, which is detailed later in Chapter 5.

Inlining is an optimization that reduces function call and return overhead by replacing a call of a function with the instructions that make up the called function. Once a function

```
for (i=0; i < NUM_BLOCKS; i++) {
    switch (check 3(i)) {
```

```
    case 3:                /* 25% */
     X[i] = 3; break;

    case 10:               /* 75% */
     X[i] = 10; break;

    default:               /* 0% */
     X[i] = 99 break; }
```

```
}
```

Figure 2.15: Basic block placement

```
cmpl       $3, %edx
je         ..B1.10  //jumps 25% of time
cmpl       $10, %edx
jne        ..B1.9
movl       $10, -4(%esp,%ebx,4)
jmp        ..B1.7  //goto loop termination


..B1.10:
movl       $3, -4(%esp,%ebx,4)
jmp        ..B1.7  //goto loop termination


..B1.9:
movl       $99, -4(%esp,%ebx,4)
jmp        ..B1.7  //goto loop termination


$B1$7: //loop termination
cmpl       $100, %ebx
jl         ..B1.4
```

Figure 2.16: Basic block example assembly listing

```
int removed(int x) {
   return x;
}
int main () {
   int y = removed(5);
   return y;
}
```

Figure 2.17: Inlining example

```
08048604 <add>:
 8048604:       8b 44 24 04          mov    0x4(%esp),%eax
 8048608:       03 44 24 08          add    0x8(%esp),%eax
 804860c:       c3                   ret
```

Figure 2.18: Correlating disassembly

is inlined, other optimizations can modify the assembly language to a larger degree resulting in difficult to observe results. In the extreme, inlining and dead code elimination can even remove substantial portions of your code such as illustrated by the code in Figure 2.17. Try compiling the code with aggressive optimization. It should be possible for the compiler to remove most of the code and return 5 with minimal instructions.

There are many other compiler optimizations that affect source code position and what is detailed here is just a small sample. Being aware of the issue and continued study on compiler optimization will improve your skills at correlating assembly code with source code.

2.4.8 Tip #7 – Correlating Disassembly to Source

The previous tip discussed correlating assembly code with source code in the face of compiler optimization. One additional level of difficulty is introduced when attempting to correlate a disassembly listing to source code because the listing does not contain a source line mapping. To help in these situations, compiling to assembly and analyzing the assembly file alongside the disassembly listing is recommended.

Consider Figure 2.18, which is the disassembly of a function named add(). To find the C source code that correlates to a particular function, search the source files in the build for the function add(). Once the file is found, the next step is to compile the source file to assembly via the -S option.

Suppose Figure 2.19 is the equivalent assembly language representation. As you can see, the assembly listing has line numbers for individual lines of assembly code. You can then use the two listings and, for example, conclude that the instruction at address 8048608 resulted from line 10 of the source file.

The above example is a somewhat simple one; large functions and libraries can complicate the analysis; however, the technique illustrated is still possible. Be aware of functions that reside in libraries that are linked into your application as you may not have source code and thus cannot compile to assembly. If you determine that a particular function does not come from the source code in your application, you can use the *nm* command to list the symbols in the libraries and then search for the particular function of interest. C++ code also adds a degree of difficulty to this technique due to *name decoration*.

2.4.9 Sample Assembly Walkthrough

This section puts the pieces together and studies the assembly language for a sample piece of code. Figure 2.20 is a listing of the sample code which includes and uses a function that computes a matrix multiplication.

```
# -- Begin   add
# mark_begin;
      .align    2,0x90
      .globl add
add:
# parameter 1: 4 + %esp
# parameter 2: 8 + %esp
..B2.1:                          # Preds ..B2.0
      movl       4(%esp), %eax                        #8.5
      addl       8(%esp), %eax                        #10.12
      ret                                             #10.12
      .align     2,0x90
# mark_end;                      # LOE
      .type      add,@function
      .size      add,.-add
      .data
# -- End add
```

Figure 2.19: Assembly listing of add() **function**

```
1    #include <stdio.h>
2    #include <time.h>
3
4    #define NUM 3
5    static double  a[NUM][NUM];
6    static double  b[NUM][NUM];
7    static double  c[NUM][NUM];
8
9    // matrix multiply routine
10   void multiply_d(double a[][NUM], double  b[][NUM],
     double  c[][NUM])
11   {
12     int i,j,k;
13     double temp = 0.0;
14     for(i=0;i<NUM;i++) {
15       for(j=0;j<NUM;j++) {
16         for(k=0;k<NUM;k++) {
17           temp = temp + a[i][k] * b[k][j];
18         }
19         c[i][j] = temp;
20         temp = 0.0;
21       }
22     }
23   }
24
25   //routine to initialize an array with data
26   void init_arr(double row, double col, double off, double a[][NUM])
27   {
28     int i,j;
29     for (i=0; i<NUM;i++) {
30       for (j=0; j<NUM;j++) {
31         a[i][j] = row*i+col*j+off;
32       }
33     }
34   }
35
36   // routine to print out contents of small arrays
37   void print_arr(char * name, double array[][NUM])
38   {
39     int i,j;
40
```

Figure 2.20: Matrix multiplication (*Continued*)

```
41      printf("\n%s\n", name);
42      for (i=0;i<NUM;i++){
43        for (j=0;j<NUM;j++) {
44          printf("%g\t",array[i][j]);
45        }
46        printf("\n");
47      }
48    }
49
50   int main()
51   {
52     // initialize the arrays with data
53     init_arr(3,-2,1,a);
54     init_arr(-2,1,3,b);
55     multiply_d(b,a,c);
56
57     print_arr("a", a);
58     print_arr("b", b);
59     print_arr("c", c);
60     return 0;
61   }
```

Figure 2.20: (*Continued*)

Appendix A includes relevant portions of the assembly listing for the program when compiled under aggressive optimization. The file was compiled with the Intel Compiler using profile-guided optimization. To mimic the steps of compilation perform:

1. Compile with profile generation
    icc -prof-gen main.c

2. Execute a.out
    ./a.out

3. Recompile with aggressive optimization
    icc -prof-use -xN -O2 -S main.c

The file, main.s, is generated and contains an assembly language listing similar to Appendix A.

Analyzing the assembly listing reveals a great amount about how the compiler optimized the source code. Focusing on the critical hot spot, the following observations about the assembly listing of the `multiply_d()` function are made:

- The start of the `multiply_d()` function occurs at the position of the `multiply_d` label (Appendix A, Figure 2, line 5). Above the label is an align directive that specifies that the text of the function should be aligned on a 2 byte boundary and to use *nop* instructions (0 × 90) to pad.

- The call of `multiply_d()` is actually to the target `multiply_d.`, which is an alternate entry for the function. Notice that the registers are already loaded with the addresses of array b, a, and c, which saves pushes to the stack and loads off of the stack common in the standard call/return sequence.

- The two branches at line 33 and line 39 represent the two outer loops. The innermost loop has been optimized by initializing a counter with a negative value at line 20 and then counting upwards. The branch at line 33 relies on *addl* to set the equal to 0 condition bit after three iterations. This optimization removes the need for an explicit compare against 0 instruction. Induction variable elimination has been performed on the outermost loop. Instead of tracking the iteration of the variable i from 0 to 3, the registers representing the correct index of c and b are incremented.

- The double precision floating-point calculations are accomplished using the SSE registers and SSE instructions on scalar double precision values. In other words, the compiler is using the SSE registers instead of the x87 registers for the calculations. One reason the SSE registers are more efficient is because x87 is stack based and requires several fswap operations between calculations.

- The innermost loop has been fully unrolled and eliminated as a result. Notice the innermost loop should iterate three times. Instead, three sets of *movsd* and *mulsd* (multiply) instructions are listed.

Chapter Summary

This chapter discussed the basic meaning of performance. Performance can include start-to-finish execution time, throughput, power utilization, or just about any characteristic that embedded systems are compared upon. This book defines performance primarily as start-to-finish execution time and will explicitly mention the other types of performance when intended.

The evolution of Embedded Intel Architecture processors has spanned over 30 years of evolution from the first microprocessor, the 4004 which targeted application as a business calculator, to successive iterations of processors with corresponding improvements in performance and integration levels. The IA-32 architecture processors are trending to cover different segments of the market with a wide range of processors from multi-core processors to low power processors and SOC integration. The history provides insight into some of the key features and capabilities added to historical and current processors.

Understanding IA-32 architecture assembly language is challenging due to the sheer number of instructions, different registers, and different methods of accessing memory. A tutorial is provided to assist you in reading the assembly language in your debug and performance work.

Related Reading

For more information on what is termed the "Power Wall" you are referred to an early prediction [3] of this limiter made in 2001.

A good resource for learning IA-32 architecture assembly language in detail is "Assembly Language For Intel-Based Computers," by Kip Irvine.

An excellent and very detailed book on the Pentium 4 processor and previous processors is authored by Tom Stanley entitled, "The Unabridged Pentium 4: IA32 Processor Genealogy."

References

[1] V. Pallipadi, *Enhanced Intel SpeedStep® Technology and Demand-Based Switching on Linux**, http://www3.intel.com/cd/ids/developer/asmo-na/eng/dc/xeon/reference/195910.htm

[2] *Intel® 64 and IA-32 Architecture Software Developer's Manuals*, www.intel.com/products/processor/manuals/index.htm

[3] P. Gelsinger, Microprocessors for the new millennium: Challenges, opportunities and new frontiers, *ISSCC Tech Digest*, 22–25, 2001.

## CHAPTER 3

Multi-core Processors and Embedded

Key Points

- Performance is the main motivation for employing multi-core processors in embedded systems; however, the demand for performance is balanced by the need to keep power utilization within reasonable limits.

- Many applications in the embedded market segments can benefit from the increased performance offered by multi-core processors.

- Industry-standard benchmarks provide single-core, multi-core, and power performance information that can help characterize embedded processors. This is a beneficial step before committing to a particular processor architecture for your embedded development project.

Increased integration has been a hallmark of processor evolution for many years. Similar to how different types of functionality have been integrated onto a single processor core, the sheer availability of transistors has made it feasible to package together several processor cores resulting in the development of multi-core processors. The move to multi-core processor architectures in the embedded processor industry has been spurred by other trends occurring over the past few years as well, including:

- Diminishing returns from *instruction level parallelism* (ILP) enhancing techniques – the microarchitectural techniques employed to extract parallelism from one instruction stream are becoming more expensive relative to the performance gain.

- Clock scaling reaching limits due to power constraints – an empirical study [1] suggests that a 1% clock speed increase results in a 3% power increase. Space, power, and fan noise of cooling are key customer constraints in many market segments.

This chapter provides a summary of multi-core processors including details on why the industry is moving to multi-core processors. A primer on multi-core processors that details basic terminology relevant in their use is included. This chapter then shifts to a review of several embedded market segments with a focus on if and how multi-core processors are being employed in each. Finally, a special section on the benchmarking of multi-core processors is included to help educate users on evaluating reported performance and on how to use *benchmarks* to effectively characterize system performance.

3.1 Motivation for Multi-core Processors

There is no conclusive study that supports my claim of diminishing returns from ILP techniques; however, it can be inferred through the trend to multi-core. The sheer fact that vendors have produced first SMT-enabled processors and now multi-core processors implies that more performance can be returned applying the available transistors to add processor cores as opposed to ILP-enhancing techniques. Figure 3.1 provides further insight into why multi-core processing is particularly attractive today. The power and performance of three different configurations running a set of applications are compared. These configurations from left-to-right are:

1. Standard single-core processor over-clocked 20%

2. Standard single-core processor

3. Dual-core processor each core under-clocked 20%

The observation for configuration 1 compared to configuration 2 is that performance on these applications is increased 1.13x with a power increase of 1.73x. This compares with configuration 3 where performance is increased 1.73x with a power increase of 1.02x. On the right class of applications, multi-core processors offer higher performance with a smaller increase in power compared to straight frequency scaling. Two caveats from this observation is that not all applications will be able to take equal advantage of multi-core processors and that more work is required from the developer to realize these performance gains.

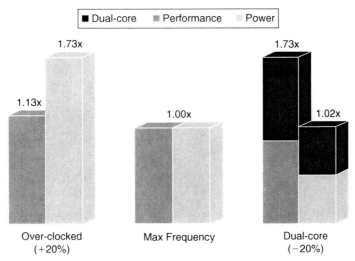

Figure 3.1: Multi-core performance and power scenario

The underlying reason why increased frequency is leading to a disproportionate increase in power is understood by reviewing the power equation in *Complementary Metal-oxide Semiconductor* (CMOS) microprocessors. A simplification of the full power equation is:

$$Power = C \times V^2 \times F$$

where C is dynamic capacitance, V is supply voltage, and F is clock frequency.

Dynamic capacitance is a function of the number of transistors in the processor so typically trends linearly. Since the contribution of voltage is squared in the equation, the supply voltage required to drive the transistors is the dominating factor in the equation. One relationship between frequency and voltage is that as you increase frequency, typically voltage must be driven up to ensure the state change of transistors is quickly recognized. Correspondingly, a drop in frequency results in the ability to drive voltage lower. As a result, you can understand why it is possible to provide two cores at a 20% reduction in frequency that use approximately the same amount of power as one core.

3.2 Multi-core Processor Architecture

Multi-core processor implies two or more processor cores in the same physical package. As mentioned in Chapter 1, systems with multiple processor cores have been available for several years in the guise of multiprocessor systems. Multi-core processors differ from multiprocessor systems physically as detailed in Figure 1.1 and differ on a number of characteristics:

- Communication latency

- Bandwidth

- Number of processors

The communication latency of a multi-core processor is typically lower than a multiprocessor. Bandwidth between cores of a multi-core processor is also typically higher than a multiprocessor. The reason is the proximity of the processor cores. Consider the processor cores in the Dual-Core Intel® Xeon® processor 5100 series, which is a multi-core multiprocessor. Two processors are connected via a high speed front side bus (FSB) and each processor contains processor cores that share the L2 cache. The latency to the L2 cache is on the order of 10–20 cycles and to main memory is on the order of 100 cycles. The data width to the L2 cache is 256 bits and provides bandwidth on the order of 64 GBs. The FSB operates at 667 MHz and provides bandwidth on the order of 5 GB per second. The latency and bandwidth between two cores on the same processor is on the order of 10 times better than between cores on different processors.

There is currently a difference between the number of processor cores available in a multi-core system and the number that have been available in multiprocessor systems. Current multi-core IA-32 architecture processors have a maximum of four cores in a package. Previous multiprocessor systems have been available with up to 64 IA-32 architecture processor cores available in one system. The expectation is the number of cores made available in multi-core processors will increase over successive generations; however, there are some fundamental hardware and software challenges that will need to be solved before large scale multi-core processors become practical.

A current example of a multi-core processor is the Intel® Core™ Duo processor, which is comprised of two similar processor cores in the same die. Figure 3.2 shows two typical multi-core processor architectures. A somewhat simple manner of envisioning a multi-core processor is as two or more processor cores connected together inside the chip packaging as opposed to a multiprocessor where multiple processor cores may be in a system, but connected via the system bus. The two multi-core architectures in Figure 3.2

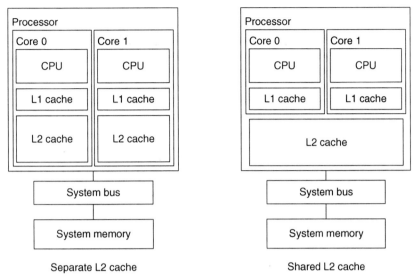

Figure 3.2: Multi-core processor with separate L2 and multi-core processor with shared L2

differ in the location of connection between the two processor cores. The multi-core processor on the left side of Figure 3.2 shows two independent processor cores, each with a level 2 cache, and sharing system bus and connection to memory. The multi-core processor on the right side of Figure 3.2 displays two independent processor cores sharing the level 2 cache, system bus, and connection to memory. There are advantages and disadvantages of the two layouts; however, the current trend is toward shared level 2 cache, which offers the following advantages:

- Size allocated per core can be dynamically adjusted with the potential of the total level 2 cache being available to applications that require it.

- Sharing of items between cores. Threads on separate cores can synchronize through the faster level 2 cache as opposed to synchronization through main memory or the next level of cache.

3.2.1 Homogeneous Multi-core and Heterogeneous Multi-core

One classification of multi-core processors is based upon the type of processor cores in the package. A *homogeneous multi-core processor* is comprised of processor cores that support the same instruction set architecture (ISA). The Intel® Core™ Duo processor is an example of a homogeneous multi-core processor where the two processor cores are identical.

A *heterogeneous multi-core processor* is comprised of processor cores that support different ISAs. A heterogeneous multi-core processor has the advantage of employing specialized processor cores designed for the task. For example, the Intel® CE 2110 Media processor consists of an Intel® Xscale processor core and an Intel® Micro Signal Architecture (MSA) DSP core. The Tolapai SOC Accelerator is a *System-on-a-Chip* design incorporating a low power IA-32 architecture processor and a series of packet processing engines. The advantage of a heterogeneous multi-core processor is the ability to employ processor cores optimized for the specific needs of the application. In the case of the Tolapai SOC Accelerator, the integrated IA-32 architecture processor handles general purpose processing as well as offering a wealth of pre-existing software to ease application development. The packet processing engines accelerate network processing functions such as cryptography, IP security, and virtual private network algorithms.

A new classification of multi-core processor is termed *many-core*. A many-core processor is comprised of a large number of processor cores, not necessarily identical, and not necessarily as full featured as the individual cores that comprise a typical multi-core processor. A hypothetical many-core processor, such as the one depicted in Figure 3.3, could consist of a few powerful superscalar IA-32 architecture processor cores suited to handle general purpose calculations and an array of in-order execution lower power IA-32 architecture processor cores that handle specialized processing.

3.2.2 Symmetric and Asymmetric Multi-core

Another classification of multi-core processors pertains to how the processor cores appear in relation to the embedded system as a whole. There are two classifications, termed symmetric and asymmetric, that differ in terms of:

- Memory access

- Operating systems

- Communication between processor cores

- Load balance and processor affinity

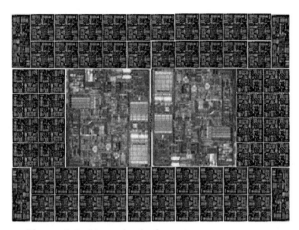

Figure 3.3: Hypothetical many-core processor

In *symmetric multiprocessing* (SMP), the processor cores in the system have similar views of the system and thus share main memory. This means that values stored in main memory can be accessed and modified by each of the processor cores. In *asymmetric multiprocessing* (AMP), the view of the system is different between processor cores. So memory regions could be disjoint, implying two processor cores cannot directly access and modify the same memory location.

Typically, an SMP will have one operating system (OS) that controls the processor cores. An embedded system configured as an AMP system may contain multiple OSes that effectively partition the system.

Communication between processor cores in an SMP system is accomplished by sharing memory and relying upon the OS to provide functionality to support synchronization and locking of memory. In AMP systems with disjoint memory regions, communication is accomplished through a message passing facility that serves to marshal data and pass it from one OS to another for use. Multicore communication application programming interface (MCAPI) [2] is an example of a message passing standard that supports communication between processor cores in an AMP system. Another technique for sharing values in an AMP system is to provide a memory map that uses specialized hardware to allow two cores to share distinct address spaces between processor cores.

Load balancing in an SMP is accomplished by the OS and subject to its scheduling. For example, an SMP implementation of Linux will treat the processor cores as a collection of work resources and schedule threads to the cores based upon the current load running on each core. SMP-enabled OSes support *processor affinity*, which allows the assignment of processes to specific processor cores. An AMP system is statically partitioned so the processor cores assigned to one OS will not be shared by a different OS on the embedded system.

3.3 Benefits of Multi-core Processors in Embedded

Multi-core processors offer developers the ability to apply more computer resources to a particular problem. These additional resources can be employed to offer two types of advantages, improved turnaround time or increasing throughput. An improved turnaround time example is the processing of a transaction in a point-of-sales (POS) system. The time it takes to process a transaction may be improved by taking advantage of multi-core processors. While one processor core is updating the terminal display, another processor core could be tasked with processing the user input. Increase in throughput can occur in the servers that handle back-end processing of the transactions arriving from the various POS terminals. By taking advantage of multi-core processors, any one server could handle a greater number of transactions in the desired response time. Balancing both improved turnaround time and increased throughput is power utilization. A power budget may be specified for the individual system used on the client end or the collection of back-end servers. The additional performance may lead to space, power, and cooling cost savings as fewer systems would be required to meet a specific level of demand.

Scalability is the degree to which an application benefits from additional processor cores; more processor cores should result in a corresponding increase in performance. Figure 3.4 shows possible performance scaling improvements derived from multi-core processors. The area labeled "Supralinear" indicates a performance improvement that scales greater than the number of processor cores. Supralinear speed-up is rare, but possible in cases where the application benefits from the collective increased cache size and layout [3]. The first line shows the theoretical case where the performance scales with the number of processor cores. The "Typical scaling" curve shows what a sample application may return performance-wise as the number of processor cores used to run the application increases. This curve displays where there is diminishing performance as the number of processor cores employed is increased. Less than linear scaling is

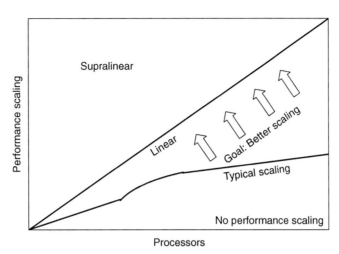

Figure 3.4: Multi-core performance scaling goal

indicative of either insufficient parallelism or communication overhead of the parallelism. Some communication overhead in a parallel application is unavoidable and so in many cases even linear speed-up is not possible. The goal of your effort to employ parallelism is to increase the amount of scaling possible in your application.

3.4 Embedded Market Segments and Multi-core Processors

Embedded Systems span a wide range of applications and market segments. Methods of employing multi-core processors in these market segments differ based upon the unique needs of the device. This section summarizes several embedded market segments and provides examples of how multi-core processors can provide benefit to applications in each segment specifically around which type of performance is needed by the application and how power utilization impacts. In general, the main driver for employing a multi-core processor design will either be performance improvement or cost benefits through consolidation.

3.4.1 Wireless Telecommunications Infrastructure

Backing each active cell phone is an array of computing platforms that transmit to and receive information from cell phones, route voice and media data across the network, and seamlessly hand off direct communication as the phone moves out of range of one cell phone tower and into range of another tower or base station. A trend in wireless telecommunications infrastructure is toward modular building block architectures that can be deployed across systems such as Radio Network Controllers (RNC), Access Service Node Gateways (ASN GW), and Base Transceiver Stations (BTS). In addition, the network communications infrastructure can also benefit from these very same components in functions such as IP Multi-media Subsystems (IMS), Server GPRS Support Nodes (SGSN), and Gateway GPRS Support Nodes (GGSN).

The primary benefit of multi-core processors in this setting is throughput – the ability to handle more calls and richer data formats. The primary challenge in taking advantage of multi-core processors for many of these applications is with the applications themselves; they tend to be large, *legacy applications* that are very difficult to multi-thread. Many of these applications execute on proprietary OSes that are not SMP enabled. Given the development cost of the substantial changes necessary to take advantage of multi-core processors, a multi-phase transition may be a reasonable approach. Such a transition strategy over several generations of the product would be:

1. Port the software to execute on an SMP-enabled OS. Take advantage of process level parallelism.

2. Incrementally add *threading*.

3.4.2 Industrial Control

Industrial control employs computers to aid in monitoring and directing factories that build a wide range of products used in the industrialized world. Efficient monitoring, *data acquisition*, and dissemination of information on the factory floor leads to more efficient production. The industrial applications targeted by multi-core processors span test and measurement instrumentation, human machine interfaces, industrial PCs, and automation control systems.

One innovative use of multi-core processors in industrial control is the consolidation of real-time control components and human machine interfaces using virtualization and multi-core processors to enable the unique demands of each of these applications. Real-time control components typically require *real-time response* and as such demand dedicated processor core resources and an OS optimized to provide this functionality. Human machine interfaces demand rich GUI interfaces common in desktop/server OSes that do not provide real-time response. Using virtualization, embedded developers can partition the cores in a multi-core processor system to provide each task with what they need [4].

TenAsys Intime* software is an example development tool that allows embedded developers to partition a machine to execute a real-time OS alongside a Windows OS* and defines an interface application programming interface (API) to transfer data between the OSes.

3.4.3 Federal (Military, Aerospace, Government)

Embedded system applications in the military, aerospace, and government segments (MAG) have two requirements above and beyond the functional needs specific to the application. First, the system must integrate well with already deployed systems and offer flexibility to work with future deployed systems. This requirement makes it desirable to employ commercial off-the-shelf (COTS) systems that enjoy inherent economies of scale. Second, these systems need to handle harsh operating environments and extended lifetimes.

Some guidance systems such as those in satellite control systems only require the performance of older processors such as the Intel® 186 processor; however, there are guidance systems being envisioned that require the performance offered by multi-core processors. Figure 3.5 is a picture of the guidance system installed on Stanley, the winner of the 2005 DARPA Grand Challenge which was a 132 mile race driven by vehicles with no human intervention. The guidance system was comprised of three *blade servers* each containing a Intel® Pentium® M processor and needing to fit within a 90 W power budget. The vehicle's speed was limited by the guidance system's ability to process terrain data. Today, systems based upon Intel® Core™ 2 Duo processors fit in the same power budget enabling the possibility of using six processor cores on an application where only three processor cores were possible a few years ago. In addition, each processor core is more powerful than the processor core used previously.

Multi-core processors offer cost effective performance to handle the tasks carried on by devices in these systems such as data acquisition, signal processing, telemetry, operator display and control.

Figure 3.5: Autonomous vehicle navigation system

3.4.4 Enterprise Infrastructure Security

Protecting corporate information networks requires a wide range of networking products that can be classified into the enterprise infrastructure security market segment. Some of the common components of these systems are firewall, virtual private network, and secure sockets layer (SSL) acceleration. Today's more sophisticated attacks on these systems is motivating integration and the ability to continually extend capabilities. For example, intrusion detection systems can be extended to detect new types of attacks and as such require either more time to analyze packet data or higher performing processors. This kind of escalation of attacks and responses can benefit from the performance benefits offered by multi-core processors.

3.4.5 In-vehicle Infotainment

Information access and entertainment in the automobile have progressed substantially from the car radio of yesteryear. With the advent of satellite and cellular-based networks, vehicles can have as much access to the Internet as home computer systems. Integrated *global positioning systems* (GPS) and media players push the requirements for processing power. In-vehicle infotainment is broadly defined as a computer system with a subset of the following functionality:

- Entertainment – DVD video and digital audio (MP3)

- Radio – standard AM/FM radio and/or satellite radio

- External communications – Wi-Fi* and WiMax* for network access

- Navigation – GPS and route plotting services

- Internal connectivity – bluetooth for cellular phone, MP3 player connection.

Current in-vehicle infotainment systems are power constrained and do not require the performance offered by a multi-core processor. Instead, power utilization and space are overriding factors. For example, the Intel® Embedded Compact Extended Form Factor reference design employs a single-core processor optimized for low power. In the future, multi-core processors will play a greater role in in-vehicle infotainment as the multi-tasking needs of future in-vehicle infotainment systems expand and multi-core processors are employed in the lowest power processors.

3.4.6 Interactive Clients

Embedded systems are being employed as a matter of convenience for many tasks that traditionally required a person or to help supplement existing staff. Examples of these interactive clients are check-in kiosks at hotels and airports, grocery checkout machines, and information kiosks at shopping malls. Demands on these types of systems are increasing with the desire to provide human-like responses, a more interactive experience, and offloading traditional server side processing onto the client. Multi-core processors offer performance increase that can be tapped by these applications. Opportunities for these types of clients abound even spurring the development of specialized OSes such as Windows Embedded for Point of Service.*

3.4.7 Voice and Converged Communications

Corporate networks are being employed to transmit more than just traditional Internet data. Voice over IP (VoIP) applications, instant messaging, and on-demand video are providing specialized capabilities that can be handled by different devices; however, maintainable and cost-effective products are being developed that integrate these features into one device, termed services over IP (SoIP). A multi-core processor-based system is well suited to the parallelism required to service these different applications in one system.

3.4.8 Digital Security Surveillance

Digital security surveillance (DSS) marries computer video processing technology with security camera systems to enable processing, analyzing, and recording large amounts of video information for criminal behavior. In these systems, automated systems can identify and track faces and classify suspicious movement for analysis by human operators. In addition, the efficient compression and recording of the data can be used and analyzed at a later point if needed.

Intel® Digital Security Surveillance Platform Technology employs multi-core processors, video processing instructions and libraries offered through the SSE instruction set and Intel® Integrated Performance Primitives, along with server and storage technology. DSS can benefit from performance in all three vectors:

1. Faster turnaround time – enables more sophisticated classification algorithms to be employed on the same data.

2. Increased throughput – enables more video feeds to be processed by a system. This capability is equally dependent on components other than the processor including I/O and storage devices.

3. Better power utilization – enables low noise, fan less operation [5] in sound critical environments such as a monitoring center.

The use of multi-core processors is performance driven providing the ability to process a greater volume of data with more sophisticated processing algorithms.

3.4.9 Storage

Richer and more varied data formats are increasing the demands for devices capable of storing terabytes of data. Customers desire access to their data from multiple locations which places additional requirements for security. In addition, the data needs to be backed up and restored efficiently in the case of a device failure.

The three classifications of storage devices used in the enterprise today are:

1. Storage Area Network (SAN) – storage devices connected to a dedicated network for storage data offering fast, reliable, and continuous data availability.

2. Network Attached Storage (NAS) – storage devices connected to LAN connection.

3. Direct Attached Storage (DAS) – one or more hard drives directly connected to system through a bus and housed in the system chassis or a directly connected device.

Multi-core processors have a role in these devices and are currently geared for the high end products in each of these storage classifications. The use of multi-core processors in the mid-range and low end products will occur as multi-core processors become the common building block across all performance segments.

3.4.10 Medical

Applications of embedded systems in the medical segment span databases to administer patient records, handheld devices for portable access to either stored information or diagnosis, and complex measurement, scanning, and treatment systems. Multi-core processors offer higher performance in a lower power envelope for these devices. Some examples of what higher performance enables include:

*   Imaging systems that once required large fixed systems can now be made portable and lower cost, enabling access to, for example, cart-based ultrasound scanners.

*   Imaging systems that merge multiple medical image sets from PET, CT, and MRI into 3D and 4D images that can be examined in real time. Faster processing can reduce patient discomfort during the scanning process [6].

3.5 Evaluating Performance of Multi-core Processors

Determining the specific multi-core processor that will suit your embedded application needs is a challenge. Relying upon the marketing collateral from a given company is not sufficient because in many cases, the results quoted are specific to a given platform and ambiguous application; there is no guarantee your application will exhibit the same performance. Therefore, it is important to understand the strengths and weaknesses of common *benchmark suites* that are used to gauge single processor and multi-core processor performance. In addition, the increased focus on energy efficiency has led to the development of benchmark suites that gauge power performance; a description of these benchmark suites is included. Finally, two practical examples of using benchmark results to estimate system and application behavior are discussed:

- How to use benchmark data to characterize application performance on specific systems.

- How to apply performance benchmark suites to assist in estimating performance of your application.

3.5.1 Single-core Performance Benchmark Suites

A number of *single-core benchmark* suites are available to assist embedded development engineers assess the performance of single-core processors. Before considering multi-core processor performance on parallel applications, scalar application performance should be reviewed. Some popular benchmark suites commonly used to evaluate single-core embedded processor performance are:

- EEMBC Benchmark Suites

- BDTI Benchmark Suites[*]

- SPEC CPU2000 and CPU2006

The Embedded Microprocessor Benchmark Consortium (EEMBC), a non-profit, industry-standard organization, develops benchmark suites comprised of algorithms and applications common to embedded market segments and categorized by application area. The current EEMBC benchmark suites covering the embedded market segments are:

- Automotive 1.1

- Consumer 1.1

- Digital Entertainment 1.0

- Networking 2.0

- Network Storage

- Office Automation 2.0

- Telecom 1.1

The EEMBC benchmark suites are well suited to estimating performance of a broad range of embedded processors and flexible in the sense that results can be obtained early in the design cycle using functional simulators to gauge performance. The benchmarks can be adapted to execute on *bare metal* embedded processors easily as well as execute on systems with COTS OSes. The classification of each application or algorithm into market segment specific benchmark suites make it easy for market specific views of performance information. For example, a processor vendor targeting a processor for the automotive market segment can choose to measure and report Automotive 1.1 benchmark suite performance numbers. Executing the benchmark suites results in the calculation of a metric that gauges the execution time performance of the embedded system. For example, the aggregate performance results from an execution of the networking benchmark suites is termed NetMark[*] and can be compared to the NetMark value obtained from an execution of the benchmark on different processors. Additionally, the suites provide code size information for each benchmark, which is useful in comparing tradeoffs made between code optimization and size. Publicly disclosed benchmark suite results require certification by EEMBC, which involves inspection and reproduction of the performance results and which lend credibility to the measurements. This is especially critical when the benchmark code is optimized to provide an implementation either in hardware, software, or both to maximize the potential of the processor subsystem. Access to the benchmark suite requires formal membership in the consortium, an academic license, or a commercial license. For further information on EEMBC, visit www.eembc.org.

The BDTI Benchmark Suites focus on digital signal processing applications, such as video processing and physical-layer communications. One valuable feature of these suites is that they are applicable to an extremely broad range of processor architectures, and therefore enable comparisons between different classes of processors. The BDTI benchmarks define the functionality and workload required to execute the benchmark, but do not dictate a particular implementation approach. The benchmark customer has the flexibility of implementing the benchmark on any type of processor, in whatever way is natural for implementing that functionality on that processor. The benchmark results developed by the customer are then independently verified and certified by BDTI. The rationale for this approach is that it is closer to the approach used by embedded developers. Embedded system developers obtain source code for key functional portions and typically modify the code for best performance either by optimizing the software (e.g., using intrinsics) or offloading some work to a coprocessor. For background on BDTI and their benchmark offerings, please consult the website, www.BDTI.com.

Standard Performance Evaluation Corporation (SPEC) CPU2006 is comprised of two components, CINT2006 and CFP2006, which focus on integer and floating point code application areas, respectively. For embedded developers, CINT2006 is more relevant than CFP2006; however, portions of CFP2006 may be applicable to embedded developers focused on C and C++ image and speech processing. CINT2006 is comprised of 9 C benchmarks, 3 C++ benchmarks and cover application areas such as compression, optimization, artificial intelligence, and software tools. System requirements are somewhat steep for an embedded multi-core processor requiring at least 1 GB of main memory and at least 8 GB of disk space. Overall SPEC CPU2006 and the recently retired SPEC CPU2000 provide good coverage of different application types. Due to the longevity of the benchmark and availability of publicly available performance data, it is possible to create a model that estimates your application performance on new processor cores before you have access to the new processor. An example of this technique is detailed later in this chapter. For background on SPEC and their benchmark offerings, please consult the website, www.spec.org.

3.5.2 Multi-core Performance Benchmarks

Good benchmarks for evaluating embedded single-core processor performance have been available for years; however, benchmarks for embedded multi-core processor performance are scarcer. Some popular benchmark programs commonly used to evaluate multi-core embedded processor performance are:

- SPEC CPU2006 (and CPU2000) rate
- EEMBC MultiBench software
- BDTI Benchmark Suites[*]
- SPEC OMP

SPEC CPU2006 evaluates multi-core processor performance when executed under what is termed "rate" mode. SPEC CPU2006 rate tests multi-core performance by executing multiple copies of the same benchmark simultaneously and determining the throughput.

The number of copies executed is determined by the tester, but is typically set equal to the number of processor cores in the system. SPEC CPU2006 rate provides a relatively straightforward method of evaluating multi-core processor performance, however, can be criticized for not representing typical workloads – how many users would execute multiple copies of the same application with the same data set simultaneously?

The EEMBC MultiBench benchmark software is a set of multi-context benchmarks based upon embedded market segment applications. Much of the work is in the creation of a platform independent harness, called the Multi-Instance-Test Harness (MITH), which can be ported to different embedded systems, some of which may not support a standard thread model such as POSIX threads or execution on a COTS OS. The group has targeted applications from several sources including creating parallel versions of the single-core EEMBC benchmarks.

The key benefit of MITH is that it allows the easy creation of a wide variety of workloads from predefined work items. As depicted in Figure 3.6 a workload can be composed of the following testing strategies:

1. A single instance of a work item – similar to the SPEC CPU2006 rate benchmark.

2. Multiple instances of the same work item, where each instance executes a different data set – provides an environment similar to embedded applications such as a multiple channel VoIP application.

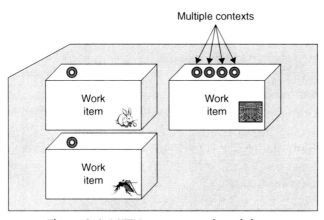

Figure 3.6: MITH contexts and work items

3. Multiple work items can be executed in parallel – emulates a complex system supporting a practically unlimited number of work items. An example of such a system would be a video conferencing application that is simultaneously running MPEG encode and decode algorithms, networking (TCP), screen updates (jpeg), and even background music (MP3).

4. Instances of multi-threaded work items – some of the algorithms are built to take advantage of concurrency to speed up processing of a single data set. For example, H.264 encoding could employ each core to process separate frames.

As mentioned earlier, the BDTI Benchmark Suites are designed to be applicable to a very wide range of processors, including single-core and multi-core processors, as well as many-core processors [7] and even FPGAs. Taking advantage of this flexibility, BDTI's benchmarks have been implemented on a number of multi-core processors beginning in 2003. In particular, the BDTI Communications Benchmark (OFDM)* and the BDTI Video Encoder and Decoder Benchmarks* are amenable to multi-core processor implementations. Consistent with its philosophy of broad benchmark applicability, BDTI does not require the use of threads or any other particular implementation approach. It is up to the benchmark implementer to decide the best approach for implementing BDTI's benchmark suites on a given target processor.

SPEC OMP [8] was released in June 2001 and is based upon applications in SPEC CPU2000 that were parallelized using OpenMP directives. The benchmark assesses the performance of SMP systems and contains two data sets, SPEC OMPM2001 and SPEC OMPL2001. SPEC OMPM2001 uses data sets that are smaller than those in SPEC OMPL2001 and is appropriate for evaluating the performance of small-scale multi-core processors and multiprocessors. SPEC OMP is not recommended for evaluating embedded application performance; the application focus of SPEC OMP is the high performance computing (HPC) market segment and floating point performance. Only 2 out of the 11 applications in OMPM2001 are written in C; the rest are written in Fortran. OpenMP fits a number of embedded application areas; however, I cannot recommend SPEC OMP as a method of benchmarking multi-core processors that will be employed in an embedded project.

3.5.3 Power Benchmarks

The embedded processor industry's continued focus on power efficiency has led to demand for benchmarks capable of evaluating processor and system power usage. Like the multi-core processor benchmarks, this area of benchmarking is somewhat less mature than single-core processor performance benchmarks. There are only two benchmark suites (EEMBC EnergyBench, and BDTI Benchmark Suites) which could be considered industry standard for the embedded market segments. Other tools and benchmarks can be employed but offer less of a fit to the embedded market segments. Four benchmark suites that are used to assess power performance in embedded systems include:

1. EEMBC* EnergyBench*

2. Battery Life Tool Kit (BLTK) [9]

3. BDTI Benchmark Suites

4. MobileMark [10]

EEMBC EnergyBench employs the single-core and multi-core EEMBC benchmarks and supplements them with energy usage measurements to provide simultaneous power and performance information. The benchmark suite requires measurement hardware comprised of a host system, a target system, a data acquisition device, shielded cable, and connector block. A number of power rails on the system board are monitored at execution time and the metrics reported include energy (Joules per iteration) and average power used (Watts). EEMBC offers a certification process and an optional Energymark* rating. EEMBC Energybench uses National Instruments LabVIEW* to display results.

During certification, the equipment is used to calculate the average amount of energy consumed for each pass over the input data set and for each benchmark being certified. Power samples at frequencies which are not aliased to the benchmark execution frequency are used to capture a sufficient number of data samples to ensure a statistically viable result, and the process is repeated multiple times to verify stability.

It is interesting to note that actual results show that power consumption varies significantly depending on the type of benchmark being run, even on embedded platforms Figure 3.7 summarizes the process to ensure stable results with regard to power measurement.

Figure 3.8 shows published EnergyBench results for an NXP* LPC3180 microcontroller. One interesting observation is made by comparing the power utilization of the processor

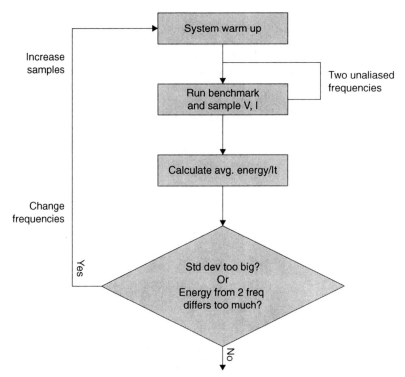

Figure 3.7: Single pass of EEMBC EnergyBench certification process

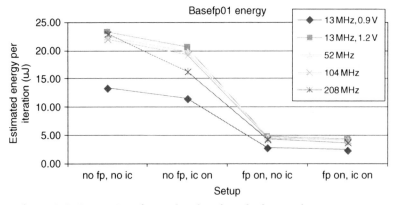

Figure 3.8: EnergyBench results showing clock speed versus power

executing at full speed versus fractions of it with and without processor features enabled such as instruction cache and floating point unit.

For example, when executing at 208 MHz with both cache and floating point unit on, the processor energy consumption for the basefp benchmark is actually in line with energy consumption of the same processor running at 13 MHz, and much better than running the processor without the cache and with the floating point unit off. Based on expected use, it may actually be more power efficient to run the processor at full speed with processor features enabled.

Figure 3.9 shows power utilization results across a number of benchmarks showing a difference of up to 10% in average power consumption depending on the actual benchmark being executed. This underlies the value of employing power utilization benchmarks to gauge performance as opposed to relying upon one "typical" figure which is often the case in designing a system today.

BLTK is a set of scripts and programs that allow you to monitor power usage on Linux systems that employ a battery. The toolkit estimates power usage by employing the

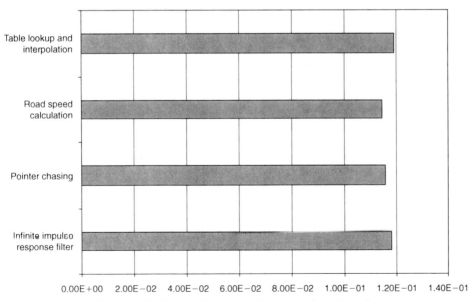

Figure 3.9: EnergyBench results across benchmarks

built-in battery instrumentation, and by running a battery from full to empty while executing a set of applications. Since the power capacity of the battery is known, it is possible to estimate average power usage taking as input the Watt hour rating of the battery and how long an application executed until the battery drains. The authors of BLTK explicitly mention that BLTK is not an industry-standard benchmark; however, the capabilities of the toolkit enable users to evaluate the impact of several embedded system design characteristics and their effect on power, such as employing hard drives with different speeds and employing more processor cores in the system.

The BDTI Benchmark suites are frequently used to estimate or measure processor energy efficiency. For estimated results, BDTI uses a consistent set of conditions and assumptions to ensure meaningful estimates. For measured results, BDTI carefully repeats processor vendors' laboratory measurements as part of its independent certification process. For example, the BDTI DSP Kernel Benchmarks[*] have been used to evaluate the energy efficiency of a wide range of single-core processors; for results, see www.BDTI.com.

MobileMark 2005 is a notebook battery life benchmark from BAPCo and is comprised of a number of workloads common in the notebook market segment including DVD playback, Internet browsing, and office productivity applications. The primary metric returned by the benchmark suite is a battery life rating. Like BLTK, the benchmark process requires conditioning of the battery and exercises the benchmarks until the battery is depleted. The rating metric is minutes. This benchmark suite requires a Windows OS which limits applicability to embedded systems employing other OSes.

3.5.4 Estimating Application Performance

One common question when customers review benchmark data is "How well will benchmark performance predict my particular application's performance if I employ your new processor or new compiler?" In other words, if a new processor or different compiler increases the performance of benchmark X by Y%, how much will the processor or compiler benefit my application? Of course, the answer is: it depends. Your application is not exactly like the benchmark program, and while the processor architects and compiler engineers may use the benchmark to assess the performance of features they are adding during development, they do not necessarily have access to or tune specifically for your application. Therefore, you should be skeptical if someone claims that you will see the same performance benefit from using a new processor or compiler as what is shown from the benchmark data. That said, you should expect some performance improvement and there are a couple of statistical techniques that can be employed to help improve the degree of confidence you have in benchmark data in estimating your end application performance. There are three techniques to use when attempting to characterize your application's potential performance improvement based upon benchmark data and they are summarized as:

- Assume the performance improvements correlate with the improvements observed in the benchmark.

- Compute the correlation between your application and a number of benchmarks and then use the performance improvement from the benchmark with highest correlation.

- Compute a multivariable regression using historical benchmark and application data to estimate a function, $f(x) = y$, where x is the performance observed on the benchmarks and y is the expected performance of your application.

The first technique and second technique are fairly easy to implement. The third technique employs a multivariable regression to estimate your application performance using the following historical data as inputs in the calculation:

- Benchmark data for a number of past processors and the new processor of interest

- Your application's performance on the same past processors for which you have benchmark data

Consider the performance data in Table 3.1. The performance results of each benchmark that comprises SPEC CINT2000 is shown for a number of processors including the Intel® Pentium® 4 processors, Intel® Core™ Duo processors, and Intel® Core™ 2 Quad processors.

The column labeled "Processor" indicates the processor on which the benchmark was executed. The column labeled "MHz" indicates the clock speed of the processor. The scores for the individual benchmark tests comprising CINT2000 are then listed.

Table 3.1: Application

Processor	MHz	164 Base	175 Base	176 Base	181 Base	186 Base
Intel Pentium 4 processor	1300	488	277	544	463	430
Intel Pentium 4 processor	1800	653	346	732	525	673
Intel Pentium 4 processor	2000	718	334	692	511	644
Dual-Core Intel Xeon processor	2800	933	935	1654	1717	1041
Dual-Core Intel Processor Xeon LV 1.67 GHz	1667	959	1128	1606	1680	1281
Intel Pentium 4 processor 631	3000	998	997	1683	1747	1099
Intel Pentium D processor 930	3000	1005	1009	1715	1785	1119
Dual-Core Intel Xeon Processor LV 2.0 GHz	2000	1147	1298	1892	1817	1537
Intel Pentium 4 processor	3800	1262	1031	2015	1415	1408
Intel Pentium 4 processor	3600	1207	1211	2046	1980	1342
Intel Dual-Core Xeon processor 5080	3733	1261	1162	2101	1788	1377
Intel Dual-Core Xeon processor 5080	3730	1282	1201	2101	1869	1415
Intel Core Duo processor T2700	2333	1338	1511	2247	2215	1888
Intel Core2 Duo processor T7400	2166	1345	1655	2578	3597	1972
Intel Core 2 Duo processor T7600	2333	1441	1782	2784	3870	2114
Intel Core2 Duo processor T7600	2333	1446	1781	2793	3852	2128
Intel Core 2 Duo processor E6600	2400	1492	1899	2931	4316	2190
Intel Core 2 Quad Extreme processor QX6700	2666	1653	2053	3209	4519	2426
Intel Core 2 Duo processor E6700	2667	1663	2062	3225	4511	2437
Intel Core 2 Duo processor E6700	2666	1654	2074	3223	4586	2432
Intel Dual-Core Xeon 5160 processor	3000	1816	2205	3070	4349	2827
Intel Core 2 Extreme processor X6800	2933	1816	2254	3525	4921	2674
Correlation		0.981	0.994	0.984	0.944	0.988

Performance estimates

197 Base	252 Base	253 Base	254 Base	255 Base	256 Base	Your app	MV estimate	Base-line
425	573	613	631	656	380	377	366	474
564	775	795	834	920	504	446	470	624
580	880	905	884	845	503	464	453	636
1156	1769	1674	1519	2419	1050	1420	1397	1381
1182	1697	1558	1412	2219	1052	1712	1727	1417
1147	1982	1753	1658	2583	1104	1493	1477	1456
1172	2002	1795	1676	2638	1113	1500	1510	1477
1387	2036	1853	1660	2589	1228	2055	2032	1663
1442	2338	2171	1996	2746	1149	1855	1847	1663
1491	2250	2136	1947	3063	1289	1805	1835	1744
1518	2355	2196	1943	2821	1301	1899	1896	1747
1441	2547	2233	2041	3035	1329	1913	1922	1793
1646	2643	2225	1966	2984	1409	2489	2489	1987
1800	2820	2693	2182	3842	1736	2493	2481	2284
1918	3034	2882	2332	4070	1850	2663	2695	2446
1933	3040	2892	2345	4096	1858	2687	2699	2455
2010	3134	3017	2515	4451	1949	2776	2773	2588
2215	3477	3310	2742	4795	2136	3075	3057	2829
2218	3488	3336	2738	4786	2150	3057	3086	2836
2218	3478	3336	2738	4858	2149	3081	3070	2844
2161	3456	3470	2878	4597	2244	3424	3424	2929
2432	3824	3666	2977	5280	2354	????	3367	3108
0.989	0.983	0.980	0.975	0.974	0.984			0.990

The "Your App" column is hypothetical data[1] and indicates the performance of your application by assuming you have measured the performance on each of the processors listed. The row labeled "Correlation" is the degree of correlation between the individual benchmark data and the data in the "Your App" column. The "MV Estimate" column was created by employing a multivariable regression where the CINT2000 benchmarks are the independent variables and the "Your App" data is the dependent variable. Finally, the baseline column indicates the overall CINT2000 rating that is part of the historical SPEC data. If you were considering moving your application from a Dual-Core Intel® Xeon 5160 processor-based system with a frequency of 3 GHz to an Intel® Core™ 2 Extreme processor X6800 with a frequency of 2.93 GHz, Table 3.2 shows a number of different performance estimates from the previous table.

The assumption in this scenario is that you have not executed the application on the Intel® Core™ 2 Extreme processor X6800 and thus would not have the "Actual" performance number (??? in Table 3.2) for it yet. The percentage difference for the "baseline" column suggests that your application would receive a 6.11% performance increase from moving to the new processor. If you applied the performance improvement from the benchmark that had the highest degree of correlation (technique #2) with your application (175 Base with .994 correlation), a 2.22% performance improvement is suggested. If you applied a multivariable regression to the data set, the performance decrease is estimated at 1.68%. The actual performance decreased by 0.99% which in this example shows the multivariable regression estimate as being the closest. There is no guarantee that employing correlation or a multivariable regression will lead to a better prediction of application performance in all cases, however, reviewing three estimates for performance improvement compared to one does provide a greater degree of confidence.

Table 3.2: Overall performance estimate comparison

	175 Base	Baseline	MV estimate	Actual
Intel Dual-Core Xeon 5160	2205	2929	3424	3424
Intel Core 2 Extreme Processor X6800	2254	3108	3367	???
Difference (%)	2.22	6.11	−1.68	−0.99

---

[1] The data is actually from the CINT2000/300 Base benchmark that serves as the data to predict in this example.

To summarize, armed with historical data from CPU2000 and your application on a number of processors, it is possible to generate two other estimates for the expected performance benefit of moving to a new processor without executing the application on the processor. The estimates based upon degree of correlation between individual CPU2000 benchmarks and on a multivariable regression may provide a more accurate indication of expected performance.

3.5.5 Characterizing Embedded System Performance

In addition to the pure performance estimates that can be created using benchmark information, benchmarks can also help characterize multi-core processor behavior on specific types of applications and across a range of design categories. For example, consider Figures 3.10 and 3.11 which depict performance of two different types of applications using varying numbers of concurrent work items on a dual-core Intel® Pentium® D processor-based system.

As you can see, on the image rotation benchmark, a supralinear performance increase is possible most likely resulting from the application's cache behavior.

On a Fixed Fourier Transforms (FFT) application using multiple work items raises the throughput by approximately 20%. Attempting to process more than two data sets concurrently results in a slight performance decrease.

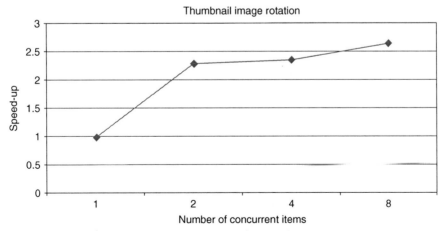

Figure 3.10: Image rotation performance

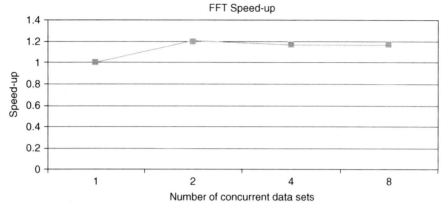

Figure 3.11: FFT performance

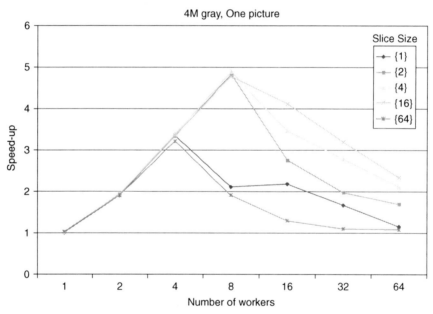

Figure 3.12: Image processing speed-up

Performance on two processor cores is the simplest multi-core case; benchmark behavior on processors with more than two cores can also be characterized. For example, consider Figures 3.12 and 3.13, which show the performance of an image rotation benchmark on a 16 core target. The figures indicate the results from running the EEMBC image rotation

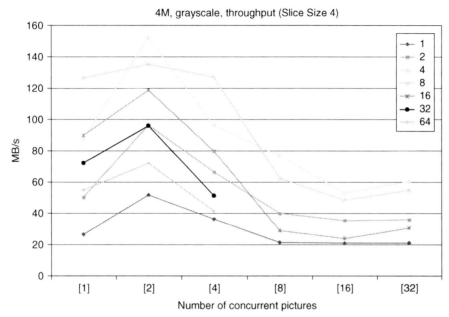

Figure 3.13: Image processing throughput

benchmark with an input data set of grayscale pictures comprised of 4 million pixels. The first graph shows the effect of using multiple cores in order to speed up processing of a single picture, whereas the second graph shows the result of trying to optimize for throughput (i.e., the total number of pictures processed). The term, slice size, refers to the granularity of synchronization that occurs between different cores working on the same picture (workers), where smaller slices require more synchronization.

Figure 3.12 illustrates that on single picture processing, using multiple cores with medium synchronization granularity, it is possible to speed up the processing by up to 5x. Note that performance scales near linearly with 2 cores active, and reasonably well with 4 cores active, but when all 16 cores are active, performance is actually slower than the 8 processor core case. Most likely, the synchronization cost when employing 16 processor cores has more of an impact than the available processing power.

Figure 3.13 shows measured throughput where each line indicates the number of processor cores working on any single picture (workers). A speed-up of 6x (better then 5x over a single image) is observed, however, this is still far from the theoretical 16x speed-up potential on a 16 processor core target. Also it is interesting to note that again, using 8 processor cores result in the best performance, split as processing 2 images at a time, with 4 cores processing each picture.

To summarize, it is the combination of multiple testing strategies and benchmarks targeting different segments that allows the analysis of a multi-core platform using multi-core benchmarks. Software developers can use the benchmarks to make intelligent decisions on structuring the concurrency within their programs, as well as evaluate which target system decisions will best serve their needs.

3.5.6 Reviewing Benchmark Data

Benchmarks can help gauge and compare embedded system performance thereby serving a useful purpose; however, benchmark results can also be abused and used to make conclusions that are not reasonable. Understanding what conclusions can be drawn depends on the characteristics of the benchmarks employed. A few helpful considerations when reviewing benchmark data include:

- System configuration
- Benchmark certification
- Benchmark run rules (base, peak)
- Reviewing single threaded and single process benchmark data.

When comparing performance results between different vendors, it is important to consider the entire system configuration and the preference is for the systems to be as similar as possible to help control for differences in these components. For example, you would not want to compare two processors using systems with different amounts of memory and disk storage and different memory latency and hard drive speed. This consideration is fairly obvious but worth mentioning.

Second, determine if benchmark results have been certified in the case of EEMBC and BDTI benchmarks or are official published results in the case of SPEC. Certified results have been reviewed by an expert third party lending the performance results greater credibility. Official published results are subject to review by other companies. Early releases of SPEC performance information on a given processor are labeled "SPEC estimate" and can be considered beta quality.

Third, consider how the benchmarks were executed. For example, SPEC allows two methods of executing the single-core benchmarks. The first measure, base results, requires the use of a limited number of options all of which must be the same for all of the benchmarks in the suite. The second measure, peak results, allows the use of different compiler options on each benchmark. It would not make sense to compare a system using base results with a system using peak results. Similarly, EEMBC allows two different types of results: out-of-the-box results and full-fury results. Out-of-the-box results specify that the EEMBC benchmark source code was not modified for execution. Full-fury results allow modification of code to use functionally equivalent algorithms to perform the tests. It may not be reasonable to compare out-of-the-box results with full-fury results unless you clearly understood the differences.

The BDTI Benchmark Suites address demanding signal processing applications; in these applications, software is usually carefully optimized for the target processor, and for this reason published BDTI Benchmark results are based on carefully and consistently optimized code. In addition, BDTI requires that vendors obtain BDTI approval when they distribute product comparisons using BDTI Benchmark Suites. As a consequence, BDTI benchmark results can be safely compared without concerns about different underlying assumptions.

Finally, scrutinize downright abuses of benchmark data. It is not reasonable to compare a single-core processor with a multi-core processor using a single-core benchmark such as CPU2000 base and claim that the multi-core processor is not doubling performance and therefore bad. For example, Table 3.1 shows that an Intel® Core™ 2 Quad Extreme processor QX6700 has a CPU2000 overall base score of 2829 and that an Intel® Core™ 2 Duo processor E6700 has a score of 2836. If you concluded that the four processor core Intel® Core™ 2 Quad Extreme processor QX6700 was therefore worse than the dual-core Intel® Core™ 2 Duo processor E6700, you would be mistaken. CPU2000 base is a single-core benchmark so should not be used to compare the benefits of a multi-core processor.

Chapter Summary

Multi-core processors are becoming mainstream in several embedded market segments due primarily to the need for increased performance while moderating the traditionally requisite increase in power utilization. This chapter detailed the motivations for the industry to employ multi-core processors, defined basic terminology, and the benefits from their use. A review of several common benchmark suites was provided to help in assessing processor vendor claims of multi-core processor performance.

Related Reading

A more detailed explanation of shared cache techniques for multi-core processors is written by Daily [11].

References

[1] P. Mehta, *Unleash the Power of Multi-Core Using a Platform Approach*, Multicore Expo, March 2006.

[2] Multicore Communication Application Programming Interface (MCAPI), www.multicore-association.com

[3] Supra-linear Packet Processing Performance with Intel® Multi-core Processors White Paper, http://www.intel.com/technology/advanced_comm/311566.htm

[4] D. Neumann, D. Kulkarni, A. Kunze, G. Rogers and E. Verplanke, Intel® Virtualization Technology in embedded and communications infrastructure applications, *Intel Technology Journal*, *10*(3), 1996, http://www.intel.com/technology/itj/2006/v10i3/5-communications/1-abstract.htm

[5] Intel® Digital Security Surveillance, *When Safety is Critical*, http://www.intel.com/design/intarch/applnots/313407.htm

[6] *How Much Performance Do You Need for 3D Medical Imaging?*, http://download.intel.com/design/embbedded/medical-solutions/315896.pdf

[7] *BDTI Releases Benchmark Results for Massively Parallel picoChip PC102*, http://www.insidedsp.com/Articles/tabid/64/articleType/ArticleView/articleId/228/Default.aspx

[8] V. Aslot, M. Domeika, R. Eigenmann, G. Gaertner, W. B. Jones and B. Parady. SPEComp: A New Benchmark Suite for Measuring Parallel Computer Performance, *Workshop on OpenMP Applications and Tools*, pp. 1–10, July 2001, http://citeseer.ist.psu.edu/aslot01specomp.html

[9] Battery Life Tool Kit, http://sourceforge.net/projects/bltk

[10] MobileMark 2005, http://www.futuremark.com/products/mobilemark2005/

[11] S. Daily, *Software Design Issues for Multi-core/Multiprocessor Systems*, http://www.embedded.com/showArticle.jhtml?articleID=183702075

## CHAPTER 4

Moving to Multi-core Intel Architecture

Key Points

- The choice of employing 32-bit or 64-bit Intel instruction set architecture (ISA) depends on your application's needs. The Intel® 64 ISA offers a larger addressable memory region and larger registers; however, it costs more in terms of data and instruction size.

- Porting legacy applications from architectures with big endian byte order may require making the code endian neutral, using binary translation software or using BiEndian Technology to deal with endianness issues.

- Understand the features of your target operating system (OS) with regard to taking advantage of multi-core processors. If your target OS does not support symmetric multiprocessing (SMP), the benefits of multi-threading will be more difficult to obtain.

- Understand the tools employed to take advantage of multi-core processors on your embedded Intel Architecture target. Compilers, performance analysis tools, multi-threaded libraries, and thread verification tools all play a role.

The desire to employ a multi-core Intel Architecture processor in your embedded system is typically motivated by performance as mentioned in previous chapters. The processor used in your last embedded system may have been different and will require you to migrate your code to execute on a new processor. The migration will present challenges in addition to the fundamental one of moving from a single-core processor to a multi-core processor. For example, the 32-bit Intel ISA contains a relatively limited number of general purpose registers. If your application is architected to use a large number of registers, modifications may be required to offer the same level of functionality and performance on Intel Architecture. Another category of issues involves data layout differences between some architectures and coding assumptions that impact the ease of migration. Understanding the cause of these issues and potential solutions are critical.

Besides fundamental architecture considerations involved in a migration, attention should also focus on the *Basic Input/Output System* (BIOS) and OS that will be employed in the embedded device. Many architectural features, both multi-core and non–multi-core relevant, require BIOS and OS support so it is important to understand the feature set, advantages, and disadvantages of the different products targeting the embedded market segments.

Once core architecture, BIOS, and OS issues are considered and addressed, development engineers engaged in a migration need an understanding of software development tools for multi-core processors. There are essentially two classes of tools for software development for multi-core processors:

- Standard tools used for single-core processor development that function differently when employed on multi-core processors.

- Specialized tools for development specific to multi-core processors.

An understanding of the capabilities and limitations of these tools will help ensure project success.

This chapter details key considerations in the migration of an application to multi-core Intel Architecture processors. The scenario or use case I envision in this chapter is a migration project from a 32- or 64-bit processor such as a PowerPC* or SPARC* to either a 32- or 64-bit Intel Architecture processor. A summary of multi-core processor-specific challenges with regard to OS, device driver, and kernel programming is provided.

The second part of this chapter discusses various tools and technology available to ease the migration to multi-core Intel Architecture processors. This high-level overview provides background on a number of important technologies including OpenMP, multi-core debuggers, and compiler support. Subsequent chapters assume the level of knowledge on these topics equivalent to what is detailed in this chapter.

4.1 Migrating to Intel Architecture

In response to the challenges of migrating to Intel Architecture, you will be required to make a set of decisions in your embedded development project that are answers to the following:

- Will you use the 32-bit x86 ISA or the Intel® 64 ISA?

- Are you impacted by endian differences and how will you resolve them?

- What BIOS and OS will be employed?

The answers to these questions require an understanding of the trade-offs involved among the many options.

4.1.1 32-Bit Versus 64-Bit Support

Migration to Intel Architecture has recently been complicated by the introduction of the Intel® 64 ISA, a new *instruction set architecture* (ISA) that increases the native data and address size to 64-bits. The 32-bit x86 ISA is the correct choice for many embedded applications; however, you may consider moving to Intel® 64 for several reasons:

- Need access to greater than 2^{32} bytes of memory.

- Primary computation is 64-bits in size.

- Embedded application is dependent on many GPRs for high performance.

4.1.1.1 32-Bit x86 ISA

The 32-bit x86 ISA has existed since the introduction of the Intel386™ processor. Many of the weaknesses of the ISA are mitigated by the software development tools and OSes so, unless you are doing system level development or assembly language programming, you possess a degree of insulation. Regardless, it is important to understand some of these characteristics so you can make the trade-off between 32-bit and 64-bit support.

A great deal of code has been developed for the 32-bit x86 ISA and it offers a well-understood and well-documented environment. From an instruction set level, the 32-bit ISA is criticized for its relative small numbers of general purpose registers, a floating point unit that has a stack-based register set and a small number of registers, and variable length instructions. Architects have recently mitigated many of these weaknesses by the use of register renaming and micro-operations in the microarchitecture and the addition of the XMM registers for SIMD floating point. If you are programming at the assembly language level you will need to understand how to take best advantage of these additions and also understand the limitations of each.

4.1.1.2 Intel® 64 Instruction Set Architecture

Chapter 2 provided a brief overview of Intel® 64 ISA and its basic register set. The key advantages of the Intel® 64 ISA over the 32-bit x86 ISA include:

- Greater number of GPRs – 16 GPRs are available instead of 8 which enable better performance as the number of temporary results that can reside in registers is increased.

- Greater number of XMM registers – 16 registers enable a higher number of temporary SIMD values to reside in registers. In addition, the 16 XMM registers can be used for scalar floating point code offering an advantage over the x87 floating point stack.

- Larger native data size – the 32-bit x86 ISA can emulate 64-bit operations albeit at a reduced performance level compared to a 64-bit processor. If your application contains many operations on 64-bit data types, you should consider using the Intel® 64 ISA.

- Greater number of byte registers – the Intel® 64 ISA features 16 registers that can be accessed as bytes compared to 8-byte registers on the 32-bit x86 ISA. If your application contains many computations on byte-sized values, the Intel® 64 ISA may offer better performance if you are programming in assembly language or use a compiler that takes advantage of the registers.

- Relative Instruction Pointer (RIP) addressing – instruction relative addressing makes position independent code easier to implement and more efficient. If you are programming or employing shared libraries, inherent performance may be improved as a result, as the compiler is able to take advantage of the addressing mode.

The key disadvantages include:

- Intel® 64 instructions use an extra byte to encode the REX prefix – the REX prefix is necessary to properly identify the instruction as an Intel® 64 instruction and can also contain source and destination register information. As a result, if your application is sensitive to code size, the larger Intel® 64 instructions may reduce performance.

- Pointer and data sizes are 64-bit instead of 32-bit – if your application is sensitive to code size, the larger Intel® 64 ISA pointers and data sizes may reduce performance. In addition, individual data items require twice as much space (64-bits versus 32-bits) in the cache, which effectively limits the number of different data values stored in cache potentially lowering cache hit rates.

Some key issues to note in migrating to Intel 64 architecture above and beyond traditional 64-bit porting include:

- LP64 versus LLP64 – this issue [1] refers to the native size of the long and pointer types in C code and these two terms designate the sizes offered by Linux and Windows respectively. LP64 specifies long and pointer types and are both 64-bit by default. The int type is 32-bits in size. LLP64 specifies long and int as 32-bit and the type long long is 64-bits. In this model, pointers are 64-bits in size as well.

- Calling convention – the calling convention differs by OS. The larger number of registers compared to 32-bit x86 ISA allows effective use of registers to pass arguments from caller to callee. On Windows, the first four arguments are passed in registers using a combination from registers RCX, RDX, R8, R9, and XMM0 through XMM3. Linux employs a more complex selection scheme for the arguments in registers drawing from RDI, RSI, RDX, RCX, R8, R9, and XMM0 through XMM7 and allow more than just four arguments to be passed. Both calling conventions require the stack to be aligned on a 16-byte boundary.

Consider the previous key critical issues in planning a migration and employing the Intel® 64 ISA.

4.1.2 Endianness: Big to Little

Endianness defines the physical ordering of component bytes that comprise larger size elements such as 16-bit, 32-bit, and 64-bit elements. Consider the 32-bit value, 0x12345678 stored at memory location 0x8000000. What specific bytes should be stored at memory location 0x80000000, 0x80000001, 0x80000002, and 0x80000003? The answer to this question is the difference between big endian and little endian format.

A big endian architecture specifies the most significant bytes coming first which implies the following layout for the bytes:

Address	0x80000000	0x80000001	0x80000002	0x80000003
Value	12	34	56	78

Little endian architecture specifies the least significant bytes at the lowest address which implies the following layout for the bytes:

Address	0x80000000	0x80000001	0x80000002	0x80000003
Value	78	56	34	12

There is no substantial advantage for architectures to layout the bytes one way or the other; this is primarily just a difference between architectures. Table 4.1 lists the byte ordering choice of several different processors. A third category of byte ordering is labeled "both" which specifies the ability of the processor to support both big endian and little endian byte ordering depending on a configuration option typically enabled in the BIOS.

Table 4.1: Endianness of various architectures

Platform	Endian architecture
ARM*	Both
DEC Alpha*	Little endian
HP PA-RISC* 8000	Both
IBM PowerPC*	Both
Intel IA-32 ISA	Little endian
Intel® Internet Exchange Architecture	Both
Itanium® Processor Family	Both
Java Virtual Machine	Big endian
MIPS*	Both
Motorola 68k	Big endian
Sun SPARC*	Big endian

4.1.2.1 Endianness Assumption

The C programming language was designed so that a program written in the language can compile and execute on any system where a C compiler exists. A typical C program that you create will compile and run on a system that contains an x86 processor or a PowerPC processor. Unfortunately, this property is not true of all C programs that can be created. One of the common causes of a C program's inability to be ported as-is in legacy applications is *endianness assumptions*. Consider the sample program in Figure 4.1.

The value *ap differs based upon the byte ordering supported by the processor. On a big endian architecture, *ap equals 0x12. On a little endian architecture, *ap equals 0x78. The code was created with an implicit assumption about where a particular byte in a larger-sized value is located. On the surface, resolving this issue does not seem difficult. A developer that was instructed to port an application from a big endian architecture to a little endian architecture would merely find code where this sort of behavior occurred and figure out how to point *ap at the correct byte. Unfortunately, this is not a robust or scalable solution. There may be a need to support both types of endian architectures so placing little endian assumptions in the code is not desirable. In addition, this example is not the only instance where an endian assumption can occur; other instances include:

- Data storage and shared memory – data that is shared between machines with different byte ordering via common storage or memory can result in incorrect reads and writes of data.

- Data transfer – data that is transferred between machines of different endian architectures will interpret a stream of bytes differently.

- Data types – access to data that is smaller in size than its declared type will reference different data elements on machines of different endian architecture.

```
#include <stdio.h>
int a = 0x12345678;
int main()
{
  char *ap = (char *) &a;
  printf("%2x %x\n", *ap, a);
  return 0;
}
```

Figure 4.1: Endianness assumption sample program

```
struct
{
  char low:4,
  high:4;
} value;

char vbyte;
value.low  = 0x1;
value.high = 0x2;
vbyte = *(char *)&value;

if (vbyte == 0x12)
{
 printf ("Big Endian\n");
}
else if (vbyte == 0x21)
{
 printf ("Little Endian\n");
}
```

Figure 4.2: Bitfield endianness code example

These assumptions can be littered throughout a large application which makes identification and removal a time consuming effort. A second example is listed in Figure 4.2.

In this example, the layout of the low and high bitfields in the struct value depends on the endian architecture and if the value is referenced as a different type, in this case, a larger type, the value differs depending on the endianness of the architecture.

Several solutions exist for endianness assumptions in code targeted for porting from one endian architecture to another. Each solution has advantages and disadvantages and it is up to the astute engineer to understand the various trade-offs intrinsic to each. Three solutions are summarized as:

- Write endian neutral code
- Employ binary translation technology
- Employ biendian compiler technology

Endian Neutral Code

Prevention is the ideal solution to endian assumptions in code. Writing code that is endian neutral is good software engineering practice. In cases where these assumptions already

```
union int_byte {
    int int_val;
    char byte[4];
}int_byte;

#ifdef LITTLE_ENDIAN
const int INT_LEAST_SIGNIFICANT_BYTE = 0;
#else
const int INT_LEAST_SIGNIFICANT_BYTE = 3;
#endif

union int_byte var;
var.int_val = 100000;

if (var.byte[INT_LEAST_SIGNIFICANT_BYTE] == 0)
    printf("The number is divisible by 256\n");
```

Figure 4.3: Endian neutral code example

exist, it is still possible to transform your code to eliminate these issues. The previous section summarized situations where an endian assumption can exist. Creating endian neutral versions of code in these cases amounts to the following:

- Data storage or shared memory – Use a common format for storing data. If you choose big endian format for data storage, use a *byte swap* macro on the little endian architecture to store the data correctly.

- Data transfer – Employ byte swap macros to translate your data to an agreed-upon network transfer byte ordering. In networking, the typical agreed-upon format is big endian.

- Data type – Data types including unions, pointers, byte arrays, and bitfields present unique challenges which are detailed more fully in the paper by Matassa [2].

One example of writing endian neutral code involving data types appears in Figure 4.3. The position of the least significant byte depends on if the code is executing on a little endian or big endian architecture. Code has been added to address the endianness assumption and to make the code endian neutral as long as LITTLE_ENDIAN is defined when compiling and executing on a little endian architecture.

QuickTransit* Technology

Transitive* QuickTransit* Technology [3] enables the use of an executable targeting one architecture on an embedded system based upon a different architecture through the use of *binary translation*. Placed in the context of our use case detailed earlier in the chapter, a binary targeted for a PowerPC or SPARC processor executes unmodified on an x86 processor-based system. At execution time, the executable is translated by taking the original PowerPC or SPARC instructions and emitting the equivalent x86 instructions into a new executable which is then executed with no appreciable difference to the user. The translation process employs an intermediate stage that allows optimizations similar to those employed in traditional compiler technology. Binary translation has performance advantages over traditional *emulation*. Traditional emulation translates each instruction one-by-one as the program is executing. If the same instruction is encountered later during execution (as would occur in a loop), the translation occurs again. In binary translation, regions of code are turned into intermediate language and recompiled during execution time, which enables optimization and subsequent caching of translated instructions. In the case of a loop in the code, a translated set of instructions would execute for the second and subsequent executions of the loop. The technique of binary translation lessens the typical performance impact associated with translating and executing a program with a different instruction set on x86. One inherent property of the technology with regard to the endianness issue is that the translation process maintains the semantic equivalent of execution on the original processor; the code in our use case would execute with big endian semantics.

BiEndian Technology

BiEndian Technology is a compiler feature that enables an application to execute with big endian semantics when executed on little endian Intel processors. At the lowest level, the technology takes advantage of the bswap instruction which converts 16-bit, 32-bit, and 64-bit values from one endian format to the other. The programmer modifies the original source code to designate the endianness of data types in the application using command line options, pragmas, and attributes. During compilation, and where required, bswap instructions are inserted at the boundary between memory and registers. In other words, if a region of code is designated to execute with big endian semantics, the data is stored in big endian format. When a load of a value into a register occurs in preparation for an operation on the value, a bswap instruction reorders the value into little endian format. If the resulting computation is then stored back to memory, a bswap instruction reorders the value into big endian format before the store. A few more details on the technology include:

- There is a performance impact to emulating big endian semantics compared to executing little endian due to the addition of the bswap instructions.

- Some functionality outside of the compiler is required to handle static and dynamic pointer initialization.

- Interfacing big endian and little endian libraries employs a novel prologue and epilogue including file mechanism to ease the number of source changes.

4.1.3 BIOS and OSes Considerations

Two of the primary decisions to make in moving to multi-core Intel architecture processors are the BIOS and OS to employ. The choice should be guided by how well the features of the BIOS and OS map to the needs of your embedded project. Take the OS decision for example which involves a natural trade-off: the more dissimilar your current OS and the OS under consideration, the more porting work will be required. Another factor in the OS decision is the established software ecosystem available on the OS. OSes with a broader user base tend to have a more diverse, competitive, and complete software ecosystem that may favor more mainstream desktop and server OSes such as Linux or Windows. Besides these two rather obvious factors (similarity to current OS and popularity of OS), a critical factor is the level of support for current and future Intel Architecture features. Processor features with benefits spanning performance, security, multi-tasking, mobility, manageability, reliability, and flexibility have been and continue to be added to Intel Architecture. Many of the features require BIOS and OS support. Some of these key features are summarized as:

- Instruction set extension support

- SMP support

- Advanced Technology (*T) support

If any of these features are needed in your embedded project, you should confirm that the features are supported by the BIOS and OS combination you plan to use. Each of these features is briefly summarized in the next few sections. Background on BIOS and current trends are also provided. Afterwards, a brief summary of feature support for a number of different OS and OS classifications is provided.

4.1.3.1 Instruction Set Extension Support

The x86 ISA has evolved over successive generations of processors from the 8-bit instruction set used in early Intel 8086 to the 32-bit ISA introduced with the Intel i386 processor and finally to the current Intel® 64 ISA. These ISA changes are relatively large scale impacting almost every aspect of the implementation. A detailed discussion of 32-bit ISA versus 64-bit ISA appears earlier in this chapter.

A middle ground of more minor extensions exists and can be as simple as the addition of the conditional move instruction that was added in the Intel® Pentium Pro processor or as complicated as SSE that includes new instructions operating on additional registers. In particular, the SSE, SSE2, SSE3, SSSE3, and SSE4 instructions require OS support to save and restore the XMM registers upon every context switch. If the OS is not aware of the XMM registers, you may be able to create code that uses the registers, but if you are employing a multi-tasking OS, your code will not function correctly. If your applications can take advantage of the features offered by these instruction extensions, you should ensure your OS supports these features adequately.

A second consideration for instruction set extension support is the use of these extensions in the OS. Does the OS take advantage of the features offered by the instruction set extensions? Many OSes provide compatibility with the lowest common denominator of feature set, which typically means a processor in the class of the Intel386 or Intel Pentium processor and limits use of instructions available in only those processors. If you are moving your application to Intel Architecture for the first time and are not constrained by the need to support older processor versions, you should look for an OS that takes advantage of the instruction set extensions. One key question to ask in this request: Has the OS been compiled using the latest compiler technology and *processor target*? If not, a good deal of performance may be left on the table.

If your current architecture offers SIMD (single instruction, multiple data) instruction set extensions similar to SSE, knowledge of both instruction sets is requisite for an effective translation. For example, PowerPC processors offer AltiVec[*] extensions which are similar to Intel x86 SSE. One helpful document if you are engaged in such a port is the AltiVec/ SSE Migration Guide [4].

4.1.3.2 Symmetric Multiprocessing (SMP) Support

SMP is defined more fully in Chapter 3. In review, SMP enables multiple processor cores to share a common memory and facilitates parallel processing. Without SMP support in your OS, it is more difficult to take advantage of one of the primary benefits provided by a multi-core processor. The OS will natively only see one processor core and forfeit the potential performance benefits offered by the other processor cores.

Employing virtualization and partitioning in your embedded application will enable some benefit to be derived from multi-core processors independent of explicit OS support; however, the ideal situation is to have SMP, virtualization, and partitioning at your disposal. Therefore, you should know if your candidate OS supports SMP and realize what you lose if it does not.

4.1.3.3 Advanced Technologies (*Ts)

New Intel processor technology features, termed "*T" (star-T), have been and continue to be added to successive generations of Intel processors and offer enhancements to security, multi-tasking, mobility, manageability, reliability, flexibility, and performance. Examples of past *Ts include Hyper-Threading Technology, Intel® Virtualization Technology, and Intel® Input/Output Acceleration Technology. Table 4.2 summarizes a few of these *T features pertinent to embedded multi-core processors.

Current and future *T features under development typically require some level of OS support. For example, Hyper-Threading Technology requires the OS to recognize that one physical processor core can actually host two logical processor cores. The OS must then

Table 4.2: Advanced technology summary

*T	Description
Hyper-Threading Technology	Allows one physical processor core to execute multiple threads in parallel.
Intel® Virtualization Technology	Allows one system to execute multiple OSes concurrently.
Intel® Input/Output Acceleration Technology	Data acceleration engine enabling higher throughput TCP applications.

initialize both logical processor cores and then handle scheduling and communication between the two logical processor cores. Effectively, the OS would need to be SMP-enabled although the performance considerations for a Hyper-Threading Technology enabled system are different than a multiprocessor with two distinct processor cores.

If your embedded project can benefit from a particular *T, employing an OS that supports the feature is essential.

4.1.3.4 Basic Input/Output System (BIOS)

BIOS is a program executed upon system startup that enables recognition and initialization of components such as processor cores, memory, and input/output devices so that an OS can boot and take control of the system. BIOS has been in use in x86 systems for many years, has become the standard means of system initialization, and has evolved to encompass other aspects of system behavior including power management, reliability features, and processor capabilities. An embedded developer has several factors to consider in choosing a BIOS to deploy:

- Support for needed feature set

- Legacy BIOS versus Extensible Firmware Interface (EFI)-based BIOS

- Independent BIOS Vendor

Like many of the discussion areas in this chapter, the BIOS should support the minimal feature set desired. For example, if SMP is required, a BIOS that enables proper recognition and initialization of the processor cores as detailed later in this chapter should be employed. Another example involves processor capabilities. In some class of applications, hardware *prefetching* has a negative impact on performance. Ensuring your BIOS supports disabling this processor feature would be required for good performance in this special case.

The second factor, Legacy BIOS versus EFI-based BIOS, concerns a trend occurring in the type of BIOS supported by Embedded Intel Architecture processors. Legacy BIOS is a classification for BIOSes implemented using older technology (assembly language, INT XX instructions to initialize various services). EFI intends to standardize the functionality of legacy BIOS and offers an extendible C software interface. UEFI 2.0 is the most recent specification and is managed by the UEFI Forum [5]. The long-term trend is toward EFI-based BIOS which includes these benefits:

- Flexibility to meet existing and future needs

- Enables extensions to support new devices using the same interfaces

- Enables faster boot times

- Provides a rich pre-boot environment to run applications for special customer needs (i.e., diagnostic, maintenance, and performance applications)

EFI was designed to coexist with legacy BIOS services allowing the system to support both EFI and legacy boot via the Compatibility Support Module. You are encouraged to choose an EFI-based BIOS in your embedded processor design.

4.1.3.5 Extensible Firmware Interface (EFI)

The EFI is a C language-based firmware alternative to traditional BIOS and emphasizes ease of use and ubiquity across platforms by defining a software interface layer between the OS and the platform firmware. Figure 4.4 shows the location of the EFI layer in the platform architecture.

The EFI specification defines a model for the interface between OSes and platform firmware. The interface consists of data tables that contain platform-related information, plus boot and run-time service calls that are available to the OS and its loader. Together, these provide a standard environment for booting an OS and running pre-boot

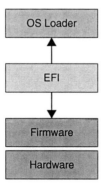

Figure 4.4: EFI layer location

Table 4.3: Independent BIOS vendors

IBV	Website
American Megatrends	www.ami.com
Insyde Technologies	www.insydetech.com
General Software	www.gensw.com
Phoenix Technologies	www.phoenix.com

applications. For more details, such as writing EFI drivers and how to use the EFI Sample Implementation and EFI Application Toolkit, see the EFI website [6].

EFI-based BIOSes are available from many independent BIOS vendors (IBVs). A sampling of IBVs that create toolkits for EFI Firmware is listed in Table 4.3.

4.1.3.6 Desktop/Server OS

The desktop and server market segments comprise a volume of the latest Intel Architecture processors much higher than the embedded market segments. In addition, the choice of OS in the desktop and server market segments is typically more constrained; Windows and Linux are the widely used OSes in these segments. In the embedded market segments, the choice of OS is varied with many different embedded OSes sharing more or less equal portions of the embedded market segment. In addition, increases in processor performance has granted Windows and Linux the ability to meet requirements traditionally met by only embedded OSes such as real-time[1] response. These new abilities coupled with the broad software ecosystem have prompted some embedded developers to consider these OSes in their embedded designs. The pure economics of the desktop and server market segments typically lead to support for the latest Intel Architecture features sooner than OSes with a smaller volume. Therefore, if support for the latest Intel Architecture features is an overriding factor in your current and future embedded projects, you may consider employing what is considered a desktop or server OS in your embedded design.

4.1.3.7 Embedded Linux OS

The popularity of Linux has led to a number of embedded OS vendors to develop and market OSes based upon the Linux distribution. These distributions are similar to the

---

[1] "Closer to real time".

Table 4.4: Embedded Linux sample

Embedded Linux	Description
MontaVista	• MontaVista Linux Carrier Grade Edition 4.0 targets telecommunications. • Professional Edition 5.0 targets several intelligent device market segments. • Mobilinux 4.1 targets wireless handsets.
Wind River	• General Purpose Platform, Linux Edition targets industrial equipment and automotive control systems. • Platform for Consumer Devices, Linux Edition targets mobile handhelds and home entertainment devices. • Platform for Network Equipment, Linux Editions targets telecommunications and features carrier grade enhancements.

desktop and server Linux versions mentioned in the previous section; however, typical Embedded Linux distributions offer features specific to the embedded market segments. These features include cross-development support, specialized and specific form factor support, and embedded specific extensions. Cross-development support is the ability to develop and debug the embedded system from a different system and OS. For example, you may have a team of developers using Windows XP as the development host and creating an embedded system that will run an embedded Linux. Specialized form factor support is the ability to tailor the Linux kernel and packages used to the minimum required for the intended application. For example, an embedded Linux employed as the OS for an MP3 player would not require full color LCD graphics support if its only output mechanism is a small text screen. Embedded specific extensions are additional capabilities added to the embedded Linux to support specific application areas. For example, carrier grade versions of Linux implement enhancements in the areas of availability, serviceability, security, and performance. Some examples of Embedded Linux OSes and a brief description of each are in Table 4.4.

The embedded Linux products mentioned in this section are royalty-free, which means cost for the product is in the form of licensing for support. There is no royalty based upon deployed production units executing the OS. Multi-core processor support is available on the current products from these OSVs.

Since embedded systems can offer unique features outside of the norm for desktop and server systems, embedded Linux distributions typically offer a board support package (BSP) required for the OS to work well with the specific embedded platform. For example, on Monta Vista Carrier Grade 4.0, the Intel® Core 2 Duo processor-based Whippany board has a different BSP than the Thermopolis board [7]. An embedded Linux would be able to offer support for the specialized components in each board if the specific BSP was used in creating your Linux target. One nice quality of these OSes is the fact that they are based upon Linux and can therefore take advantage of the same development community that exists on Desktop/Server Linux while providing customization features necessary for embedded devices. For example, support for the latest processor extensions in the Linux kernel would be available when the support is added to the open source kernel and an embedded OSV updates its product to offer the feature.

One more recent feature I have seen added as a result of customers desiring to move from a different architecture and OS combination to Embedded Linux on x86 is microstate accounting. This feature allows finer granularity timer resolution for tracking processes above the typical user, system, and real time tracked by the `time` command. The embedded OSV responded quickly to the customer which highlights one potential advantage of Embedded Linux over more general distributions.

4.1.3.8 Embedded Windows

The popularity of Microsoft Windows OSes in embedded environments has grown in recent years due to the following:

- An industry-standard interface and the wide variety of built-in functionality

- A broad software ecosystem with experienced developers and abundant applications and drivers

- Reliable support and the addition of a long-life roadmap

Microsoft has developed several embedded OSes based upon Windows that target the embedded market segments. The OSes offer a smaller footprint than the desktop and server versions and customization of the OS for the application and specific processor. Windows XP Embedded offers the ability to pick and choose specific components of standard Windows XP for use in an embedded device. Windows XP Embedded for

Point of Sales is similar to Windows XPe, but provides customized features for creating point-of-sales devices. One benefit of these OSes is the compatibility of many existing libraries and applications that run on Windows XP. For example, Windows XPe and WEPOS offer support for multi-core that is "inherited" from its sibling, Windows XP.

Windows CE is more of a "pure" embedded OS targeting application areas such as media players and handheld devices and offering support on several different processors, not just x86 processors. BSPs are available for multiple targets [8]. At this time, Windows CE does not support taking advantage of multi-core processors. Source code is available to Windows CE developers to aid in developing and debugging embedded applications under a special licensing program.

Embedded Windows With a Real-time Extension

Determinism is still a major challenge in embedded designs; although improved performance can dramatically increase processing throughput and average response times, a non-deterministic system cannot become deterministic in this manner, nor guarantee to improve worst-case response times. Some companies will identify their need as *soft real-time*, others *hard real-time*. Technically, it is either real time or it is not.

Due to the added complexity and cost of maintaining a heterogeneous computing environment, companies are striving to use only one OS, such as Windows, at all levels of an organization.

There is also an impetus to use it in other environments such as the factory floor, medical devices, simulation, test, and communications equipment. A common characteristic of these environments is that they often require hard real-time system behavior. Traditionally, developers would resort to adding a second real-time device to their system. This presents several obvious drawbacks, such as higher hardware cost, a more costly and limited integration, and higher software costs where developers are required to know how to develop for two different systems. Considering the current compute power available with multi-core processors, there is an opportunity for a real-time extension added to Windows XP. Ardence Embedded [9], now part of Citrix, provides such an extension called RTX*.

RTX is implemented as a collection of libraries (both static and dynamic), a real-time subsystem (RTSS) realized as a Windows kernel device driver, and an extended Hardware

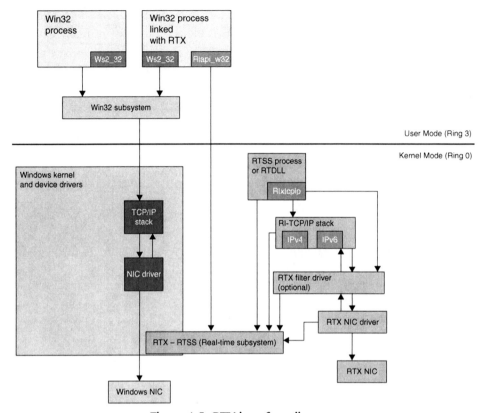

Figure 4.5: RTX interface diagram

Abstraction Layer (HAL) as depicted in Figure 4.5. The subsystem implements the real-time objects and scheduler previously mentioned. The libraries provide access to the subsystem via a real-time API, known as RtWinAPI. This RtWinAPI provides access to these objects and is callable from the standard Win32 environment as well as from within RTSS.

In addition to interrupt management and fast-timer services, the real-time HAL also provides Windows shutdown management. An RTSS application can attach a Windows shutdown handler for cases where Windows performs an orderly shutdown or even crashes. An orderly shutdown allows RTSS to continue unimpaired and resumes when all RTSS shutdown handlers return.

Before 1999, RTX executed on single processor systems only, but now supports both multiprocessor systems and multi-core systems. On a multiprocessor system, RTX can be configured in one of the following ways:

- Shared Mode – one processor core handles both RTX and Windows processing; all other cores are dedicated to Windows processing.

- Dedicated Mode – RTSS dedicates one processor core of the system to running RTSS threads, while the remaining cores run Microsoft Windows threads. This dramatically lessens the latency of real-time threads while preventing the processor starvation of Windows threads possible on a single processor system.

In embedded systems that employ multi-core processors combined with Ardence's deterministic extension to Windows (RTX), critical applications can be dedicated to an entire core, harnessing its full potential, with no interference from the OS or other applications. Benefits include performance optimization with reduced latencies and improved reliability.

The next major version of RTX plans to include SMP support. Multiple cores can be dedicated to RTX, allowing parallelism within the real-time part of the application.

4.1.3.9 Embedded OS

The recent past has seen a number of embedded device manufacturers either basing systems on desktop/server OSes or embedded versions of Linux; however, there is still a strong pull for what are termed traditional embedded OSes. Examples of this type of embedded OS include Wind River VxWorks and QNX Neutrino. These OSes typically offer real-time response and specialized support for embedded devices and specific embedded platforms in the form of BSPs. Embedded OSes offer a higher degree of customization for your embedded application than desktop/server OSes and the embedded variants of Linux and Windows.

An example of a traditional embedded OS is QNX Neutrino which features:

- SMP support

- Bound multiprocessing (BMP) support

- Extensive POSIX support

- Extended memory support

- Multiple host support (Windows XP, Solaris, Linux, Neutrino)

Bound multiprocessing support allows a mix of SMP and AMP where some applications can be bound to execute on specific processor cores while other applications can share the processor resources. This feature can also be used to preserve correct operations of software that do not work properly on multi-core processors. QNX complies with POSIX 1003.1-2001 and supports the POSIX threads API which enables explicit threading on multi-core processors. Extended memory support above 2 GB is available; however, this support is not in the form of Intel® 64 ISA support. QNX can be hosted by a number of OSes including Windows XP, Solaris, Linux, and Neutrino. BSPs are also available for several targets [10].

The Neutrino OS is supported by a set of multi-core visualization tools that provides a detailed view of the software behavior on a multi-core system. It is important to understand system level behavior when optimizing software for multi-core systems. Figure 4.6 shows the visualization tools available in QNX Momentics that consolidate system level information from all processors in a multi-core system. The QNX Momentics system profiler tool provides in-depth views such as overall CPU usage for each processor core, process and thread CPU usage, thread timing diagrams, and thread scheduling information including migration between cores.

Wind River is also enabling the use of multi-core processors having announced VxWorks® 6.6 with SMP support for several targets.

4.1.3.10 Proprietary OS

Lastly, many companies employ proprietary OSes in their embedded systems that are developed and maintained in house. This class of OS offers ultimate customization for your embedded application; however, it requires the largest amount of effort – everything you want in your OS you must implement. Likewise, if you value support for features such as SSE extensions, SMP, and Virtualization, you will need to implement support yourself.

One interesting development I have witnessed with companies employing proprietary OSes is the hosting of these OSes on other more generally available OSes. Many legacy embedded applications are either dependent on features that are no longer supported by

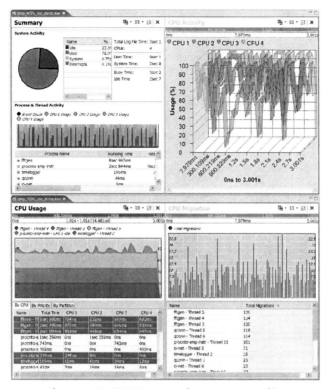

Figure 4.6: QNX Momentics system profiler

more recent OSes or are so tied into the OS that porting to a newer OS would require reimplementation of support outside of the realm of feasibility. With the increases in processor performance, it is now possible to run these proprietary OSes as an application on OSes that have support for the latest Intel architecture features by creating a shim layer that presents a view of the system that the proprietary OS expects to see. This form of virtualization will become more commonplace as multi-core processors increase in popularity and virtualization technology becomes mainstream.

4.2 Enabling an SMP OS

One fundamental piece of support for multi-core processors is OS support. In the embedded market segments, OS support is not a given due to the number of proprietary and legacy OSes employed; not everyone uses Microsoft Windows or Linux. This section provides an overview of what is required to recognize and support multi-core processors in an SMP OS. In this section, please note that a hardware thread is a distinctly different entity compared to a software thread, which is what is implied in previous sections when discussing multi-threading.

Chapter 7 of the IA-32 and Intel 64 Architecture Software Development manual includes details on Multiple-processor management [11]. The fundamental requirements for an embedded system that supports multi-core processors are:

- Process for booting and identifying processor cores

- Clear delineation between shared versus replicated resources

- Interrupt handling and distribution

- Memory synchronization and ordering

The Multiple-processor management process for Intel processors attempts to keep the bootup process fundamentally the same regardless if the processor cores are comprised of multiprocessors, SMT processors, or multi-core processors. Upon bootup, CPU identification [12] instructions are employed to determine the configuration of the processors and processor cores. Employing CPUID functionality to determine the configuration is preferred over determination based upon the explicit name of the processor. During bootup, built-in system logic appoints one processor core as the bootstrap processor and the other processor cores as application processors (APs). The bootstrap processor executes the startup code, configures the environment for handling interrupts, determines the number of processor cores, and then sends a start initialization signal to the APs. The APs initialize by executing code stored in the BIOS and then go into a halt state awaiting further instruction from the bootstrap processor.

Understanding what system structures are shared and which are replicated in a multi-core processor is important. Table 4.5 lists the various system elements and their quality with respect to being shared between processor cores or unique to each processor core.

An OS must support saving and restoring the state of the general purpose, x87 floating point, MMX, and XMM registers on context switches for every processor core assuming all of the registers are supported by the OS. This functionality makes it possible to schedule threads, interrupt them, and then resume execution at a subsequent point in time. *Memory type range registers* (MTRRs) specify the attributes of physical memory such

Table 4.5: Multi-core processor shared and unique architecture features

Architecture feature	Shared or Unique
General purpose registers	Unique
x87 floating, MMX, XMM registers	Unique
System registers (control, system table pointer, and debug)	Unique
IA32_MISC_ENABLE MSR	Shared
Memory type range registers	Shared
Performance monitoring counters	Implementation dependent

as whether or not a memory range can be stored in cache. Since memory is shared in an SMP, it makes sense that the MTRRs are shared. The IA32_MISC_ENABLE Model Specific Register specifies miscellaneous capabilities in the processor cores and is a shared resource mainly for simplicity.

Interrupts are a prerequisite for *preemptible operating systems* and are common features of modern microprocessors. Interrupts allow a processor's current stream of execution to be disrupted so that a higher priority system event can be handled. SMP OSes use interrupts as the basis for scheduling tasks on processor cores. Upon bootup, a method of setting up the interrupt structure required by the OS is employed where each processor core is assigned a unique identification which is then stored so that interrupts coming in from the I/O system are handled by the correct processor core and interrupt handler.

Memory is shared in an SMP and requires support to keep memory consistent between all of the processor cores. The challenge is how to keep memory consistent between cores while allowing the use of high performance features such as caches. A *lock* is the core functionality used to keep memory consistent by enabling a processor core to restrict access to a memory location.

Caches present an extra challenge for SMP because there can be multiple copies of a memory location residing in different caches. Cache enables fast access; however, as soon as one of the processor cores changes the value stored there then all of the cached copies of the value in the other processor cores need to be updated. In other words, a write of a shared memory value that is performed in the cache of one processor core needs to be propagated to the other processor cores. The question to consider is under what conditions can this property result in lower performance and what can be done software-wise to avoid them?

4.2.1 Basic MESI Protocol

The MESI Protocol is one of the basic cache coherency protocols employed in multiprocessing. The goal of cache coherency is to keep memory that may be cached in different processor cores consistent. As far as software stability is concerned, developers do not need to worry about this protocol other than being aware that the processor cores have a feature to maintain consistency. For performance work, it is important to understand the basics of cache coherency because effective optimization for cache requires it.

MESI stands for the four states that a particular cache line can be labeled by the cache consistency algorithm. All lines in the cache have an associated two-bit value that denotes one of the four states, which are:

- Modified – this cache line contains a copy of the memory and has modified it.

- Exclusive – this cache line contains the only copy of the memory. No other cache contains the memory. The memory has not been modified.

- Shared – this cache line contains a copy of the memory and so does at least one other cache. The memory has not been modified by any of the caches.

- Invalid – this cache line contains a copy of memory that is invalid. Either no memory is cached yet or the cached memory has been modified in a different cache.

Consider two processor cores sharing a common memory and distinct caches. For simplicity, only one level of cache is assumed and no other data values stored in the particular cache line are affected during the scenario. The following step-by-step scenario illustrates the basics of the coherence protocol and points out the performance benefits:

1. Initial state – memory contains a value and no cache has a copy of the value.

2. Processor core one references the value. Value is brought into core one's cache and the coherence state associated with the cache line containing the value is set to the Exclusive state.

3. Processor core two references the value. Value is brought into core two's cache and the coherence state associated with the cache line containing the value is set to the Shared state. In core one's cache, the coherence state associated with the cache line containing the value is set to the Shared state.

4. Processor core two modifies the value. In core two's cache, the coherence state associated with the cache line containing the value is set to the Modified state. In core one's cache, the coherence state associated with the cache line containing the value is set to the Invalid state.

5. Processor core one reads the value. The value in processor core two's cache is written back to memory. Core one loads the value and the coherence state associated with the cache line containing the value is set to the Shared state. In core two the coherence state associated with the cache line containing the value is set to the Shared state.

Notice at Step 3 where both caches have a copy of the value. As long as the value is only read from both caches, no communication between caches or to main memory is necessary, which results in good performance compared to synchronizing memory on every access. Notice at Step 4 where core two modifies its copy of the value and Step 5 where the value is then referenced by core one. The value will be synchronized with main memory and then copied into the cache of core one. The last step of the scenario details a pathological case of cache access informally termed cache ping-ponging where a value is moved back and forth between two caches. There are a number of data access scenarios that can lead to poor performance from different cache topologies.

A slight variation on the cache ping-ponging issue can occur even when a memory value is not being shared. If two memory values reside in the same cache line and two processor cores are repeatedly modifying one value, the problem of *false sharing* can occur because the cache line is the unit of granularity that is held coherent between processor cores and effectively "ping-ponging" of the cache line can occur. The general solution to false sharing is to make sure the two data values are not in the same cache line.

4.2.2 Device Driver and Kernel Programming Considerations

Embedded software developers who develop code at the system level in the form of device drivers or OS kernel modifications have special considerations when applying multi-core processors. These considerations have existed previously in multiprocessing systems, but of course the widespread employment of multi-core processors in embedded designs make them even more relevant. As hinted at previously, concurrency can lead to race conditions if not properly handled. Concurrency issues can exist in device drivers in the following cases and must be addressed appropriately:

- Basic concurrency issues – two processors execute driver code updating the same status or control register. Unsynchronized access can lead to race conditions.

- User-space access to data – a simple example is a call to printf() from two different threads, which can lead to interspersed output.

- Kernel code is preemptible – interruption of kernel code by higher priority events can corrupt unprotected work in progress.

- Device interrupts are asynchronous events – interrupts may disrupt other kernel code with in progress results thus corrupting the data values. Since the event is asynchronous, these events can be difficult to debug.

In each of these cases, designing for concurrency in the driver and kernel code is a necessity when employing multi-core processors. Solutions to *data race* issues involve synchronization employing a *critical section* around code that modifies shared data or protecting individual data with locks. A second technique is to make reentrant the driver functions and the functions they call.

A number of general guidelines when writing device drivers or kernel code are:

- Design for concurrency up front.
  - Analyze lock use for ordering. Nested locks present unique challenges.
  - Consider scalability and granularity. The region of code protected via locks or critical sections can impact performance.

- Create critical sections to serialize access to shared resources.
  - Be aware of the different techniques for creating critical sections in an OS and understand the constraints on the usage of them.

- For shared resources that do not need critical sections, declare them volatile to prevent compiler optimizations from causing problems.

- Volatile declarations are not a cure-all as the declarations have no impact on processor reordering.
- Volatile declarations do not guarantee all processors in multiprocessor system see accesses in the same order.

- Be aware of caching issues.
  - A single cache line should not contain data structures protected by more than one lock.

- Disable interrupts.
  - Sometimes required (e.g., when driver code and interrupt handler need "write" access to a data structure – interrupt handler cannot block).
  - Must be done to the smallest granularity possible.

- Design for and test your software on an SMP and preemption enabled kernel.

  - Even if this type of kernel is not your immediate target, the investment now will pay dividends if you decide to later employ multi-core processors.

- Minimize the use of non-reentrant functions where possible.

4.3 Tools for Multi-core Processor Development

Now that you have an understanding of key software ecosystem issues to be considered in moving to multi-core x86 processors, key software tools that aid in taking advantage of multi-core x86 processors are described. Compiler features supporting multi-core development are next described and include OpenMP, Automatic Parallellization, and Speculative Precomputation. An overview of thread libraries, debuggers, performance analysis tools, and graphical design tools is also provided.

4.3.1 OpenMP

OpenMP [13] is a set of compiler directives, library routines, and environment variables which is used to specify shared-memory concurrency in Fortran, C, and C++ programs. The OpenMP Architecture Review Board (ARB), which oversees the specification of OpenMP, is made up of many different commercial and academic institutions. The rationale behind the development of OpenMP was to create a portable and unified standard of shared-memory parallelism. OpenMP was first introduced in November 1997 with a specification for Fortran, while in the following year, a specification for C/C++ was released. At the time of this writing, the draft specification for OpenMP 3.0 was publicly available for comment.

All major compilers support the OpenMP language. This includes, for Windows, the Microsoft Visual C/C++. NET and, for Linux, the GNU gcc compiler. The Intel C/C++ compilers, for both Windows and Linux, also support OpenMP.

OpenMP directives demarcate code that can be executed in parallel, called parallel regions, and control how code is assigned to threads. The threads in code containing OpenMP operate under the fork-join model. As the main thread executes the application, when a parallel region is encountered, a team of threads are forked off and begin executing the code within the parallel region. At the end of the parallel region, the threads within the team wait until all other threads in the team have finished running the code in the parallel region before being "joined" and the main thread resumes execution of the code from the statement following the parallel region. That is, an implicit barrier is setup at the end of all parallel regions (and most other regions defined by OpenMP). Of course, due to the high overhead of creating and destroying threads, quality compilers will only create the team of threads when the first parallel region is encountered, put the team to sleep at the join operation, and wake the threads for subsequent forks.

For C/C++, OpenMP uses pragmas as directives. All OpenMP pragmas have the same prefix of `#pragma omp`. This is followed by an OpenMP construct and one or more optional clauses to modify the construct. To define a parallel region with a code, use the `parallel` construct:

```
#pragma omp parallel
```

The single statement or block of code enclosed within curly braces that follow this pragma, when encountered during execution, will fork a team of threads, execute all of the statements within the parallel region on each thread, and join the threads after the last statement in the region.

In many applications, a large number of independent operations are found in loops. Rather than have each thread execute the entire set of loop iterations, parallel speedup of loop execution can be accomplished if the loop iterations were split up and assigned to unique threads. This kind of operation is known as worksharing in OpenMP.

The `parallel for` construct will initiate a new parallel region around the single for-loop following the pragma:

```
#pragma omp parallel for
```

The iterations of the loop will be divided amongst the threads of the team. Upon completion of the assigned iterations, threads will sit at the implicit barrier that is at the end of all parallel regions. It is possible to split up the combined `parallel for` construct into two pragmas: the `parallel` construct and the `for` construct, which must be lexically contained within a parallel region. This separation would be used when there was parallel work for the thread team other than the iterations of the loop. A set of `schedule` clauses can be attached to the `for` construct to control how iterations are assigned to threads. The `static` schedule will divide iterations into blocks and distribute the blocks among threads before execution of the loop iterations begin execution; round robin scheduling is used if there are more blocks than threads. The `dynamic` schedule will assign one block of iterations per thread in the team; as threads finish the previous set of iterations, a new block is assigned until all blocks have been distributed. There is an optional `chunk` argument for both `static` and `dynamic` scheduling that controls the number of iterations per block. A third scheduling method, `guided`, distributes blocks of iterations like `dynamic`, but the sizes of the blocks decrease for each successive block assigned; the optional `chunk` argument for `guided` is the smallest number of iterations that will ever be assigned in a block.

By default, almost all variables in an OpenMP threaded program are shared between threads. The exceptions to this shared access rule are the loop index variable associated with a worksharing construct (each thread must have its own copy in order to correctly iterate through the assigned set of iterations), variables declared within a parallel region or declared within a function that is called from within a parallel region, and any other variable that is placed on the thread's stack (e.g., function parameters). For C/C++, if nested loops are used within a worksharing construct, only the loop index variable immediately preceding the construct will automatically be made private to each thread. If other variables are needed to be private to threads, such as the loop index variables for nested loops, the `private` clause can be added to many directives with a list of variables that will generate separate copies of the variables in the list for each thread. The initial value of variables that are used within the `private` clause will be undefined and must be assigned a value before they are read within the region of use.

The firstprivate clause will create a private copy of the variables listed, but will initialize them with the value of the shared copy; if the value of a private variable is needed outside a region, the lastprivate clause will create a private copy (with an undefined value) of the variables listed, but, at the end of the region, will assign to the shared copy of the variables the value that would have been last assigned to the variables in a serial execution (typically, this is the value assigned during the last iteration of a loop). Variables can appear in both a firstprivate (to initialize) and lastprivate (to carry a final value out of the region) clause on the same region.

In cases where variables must remain shared by all threads, but updates must be performed to those variables in parallel, OpenMP has synchronization constructs that can ensure mutual exclusion to the critical regions of code where those shared resources are accessed. The critical construct acts like a lock around a critical region. Only one thread may execute within a protected critical region at a time. Other threads wishing to have access to the critical region must wait until the critical region is empty.

OpenMP also has an atomic construct to ensure that statements will be executed in an atomic, uninterruptible manner. There is a restriction on what types of statements may be used with the atomic construct and, unlike the critical construct that can protect a block of statements, the atomic construct can only protect a single statement. The single and master constructs will control execution of statements within a parallel region so that one thread only will execute those statements (as opposed to allowing only one thread at a time). The former will use the first thread that encounters the construct, while the latter will allow only the master thread (the thread that executes outside of the parallel regions) to execute the protected code.

A common computation is to summarize or reduce a large collection of data to a single value. For example, the sum of the data items or the maximum or minimum of the data set. To perform such a data reduction in parallel, you must allocate private copies of a temporary reduction variable to each thread, use this local copy to hold the partial results of the subset of operations assigned to the thread, then update the global copy of the reduction variable with each thread's local copy (making sure that only one thread at a time is allowed to update the shared variable). OpenMP provides a clause to denote such reduction operations and handle the details of the parallel reduction. This is accomplished with the reduction clause. This clause requires two things: an associative and commutative operation for combining data and a list of reduction variables. Each thread within the parallel team will receive a private copy of the reduction variables to use when

```
static long num_rect = 100000;
double width, area;
void main()
{   int i;
    double mid, height;
    width = 1.0/(double)num_rect;
#pragma omp parallel for private(mid) reduction(+:height)
    for (i=0; i<num_rect; i++)
    {
        mid = (i+0.5)*width;
        height += 4.0/(1.0+mid*mid);
    }
    area = width*height;
    printf("Pi = %f\n",area);
}
```

Figure 4.7: Sample OpenMP program

executing the assigned computations. These private variables will be initialized with a value that depends on the reduction operation. At the end of the region with the reduction clause, all local copies are combined using the operation noted in the clause and the result is stored in the shared copy of the variable.

Figure 4.7 is a small application that computes an approximation of the value for Pi using numerical integration and the midpoint rectangle rule. The code divides the integration range into num_rect intervals and computes the functional value of $4.0/(1 + x^2)$ for the midpoint of each interval (rectangle). The functional values (height) are summed up and multiplied by the width of the intervals in order to approximate the area under the curve of the function.

A pragma to parallelize the computations within the iterations of the for-loop has been added. When compiled with an OpenMP-compliant compiler, code will be inserted to spawn a team of threads, give a private copy of the mid, i, and height variables to each thread, divide up the iterations of the loop between the threads, and finally, when the threads are done with the assigned computations, combine the values stored in all the local copies of height into the shared version. This shared copy of height will be used to compute the Pi approximation when multiplied by the width of the intervals.

OpenMP can also handle task parallel threading. This is done using the `parallel sections` worksharing construct. Parallel sections are independent blocks of code that are non-iterative. Within the sections construct, tasks are delineated by inserting a `section` construct before a block of code. These blocks of code, enclosed within curly braces if there is more than one line in the block, will be assigned to threads within the team. For example, in a three-dimensional computation, the calculation of the X, Y, and Z components could be done by calling three separate functions. To parallelize these three function calls, a `parallel sections` pragma is used to create a parallel region that will divide the enclosed block of code into independent tasks. Above each function call, a `section` construct will set the execution of each function as a task that will be assigned to one of the threads within the team.

The OpenMP specification includes a set of environment variables and API functions in order to give the programmer more control over how the application will execute. Perhaps the most useful environment variable is OMP_NUM_THREADS that will set the number of threads to be used for the team in each parallel region. The corresponding API function to set the number of threads is `omp_set_num_threads()`. This function takes an integer parameter and will use that number of threads in the team for the next parallel region encountered. If neither of these methods is used to set the number of threads within a team, the default number will be used. This default is implementation-dependent, but is most likely the number of cores available on the system at run-time.

The OpenMP specification contains many more directives, environment variables, and API. Consult the specification document for full details.

4.3.2 Automatic Parallelization

Having the compiler analyze source code and be able to determine if that code can be executed concurrently has been a research topic for many decades. Unfortunately, there have not been too many breakthroughs in this field of study beyond simple loops that can be proven to have completely independent iterations.

Dependence analysis is a static method to determine what dependencies exist between variables referenced within the loop body across iterations of the loop. If no cross-iteration data races can be shown within the loop, the iterations can be executed concurrently and the loop can be parallelized. Use of pointers within the loop and the potential of pointer aliasing can quickly defeat static dependence analysis.

If a compiler supports automatic parallelization (autoparallelization), it should have a means to print a report of what loops were considered for parallelization, the success of the attempt, and, in the case of failure, a list of the reasons that the loop was disqualified. The most useful feature of such a report is the list of the dependencies found within the loop and what variables were involved. The programmer should then be able to analyze the reasons given by the compiler and the code of the loop to determine if the dependencies are valid or if the compiler has been overly conservative in its analysis. In the case of valid dependencies, the code of the loop might be able to be rewritten to eliminate these dependencies. In the latter case, the programmer may employ pragmas or other compiler directives to instruct the compiler to ignore the imaginary dependencies.

If the programmer is disinclined to allow the compiler to thread loop iterations, autoparallelization reports can be used to analyze candidate loops for advice on using OpenMP. If no dependencies are found, the loop is simply parallelized with the `parallel for` construct. If the compiler does identify potential dependencies, the report indicates what variables need some form of protection with the concurrent execution. This can be done either by adding appropriate `private` or `reduction` clauses or the insertion of `critical` or `atomic` constructs to enforce mutually exclusive access on the variables identified.

4.3.3 Speculative Precomputation

While processor clock speeds have increased, the speed of memory access and bus speeds have not kept up the same pace. The need to have data ready for multiple threads is very evident. If an application is not threaded, there is still the need to keep relevant data available by avoiding cache misses during execution. Speculative precomputation [14] is a technique that proposes to make use of the unused cores to improve serial application performance. This performance boost would be realized by spawning speculative threads that attempt to pre-load cache in advance of the needs of the non-speculative thread.

The target of the speculative threads would be those memory loads that cannot easily be detected by prefetch hardware within the processor; that is, loads that have no predictable access patterns or chain of dependent loads. Speculative threads are spawned to look ahead of the code running in the non-speculative threads, compute the addresses of near future memory accesses, and load the data into cache. Spawning of speculative threads can be triggered when executing specific instructions within the non-speculative stream or from speculative threads spawning new speculative threads. Compiler assistance is needed to determine potential sections of the code that may exhibit access behavior that could be alleviated by speculative computation and additional threads.

4.3.4 Thread Libraries

Thread libraries represent a second set of programming tools to take advantage of threads on multi-core processors. Three different types of libraries are summarized in the next sections including explicit threading libraries, Intel® Threading Building Blocks (TBBs) [15], and domain-specific threading libraries. In addition, status is shared on the threading models currently under consideration by the C++ standard committee.

4.3.4.1 Pthreads and Win32 Threads

Explicit threading libraries require the programmer to control all aspects of threads, from creating threads and associating threads to functions to synchronizing and controlling the interactions between threads and shared resources. The two most prominent threading libraries in use today are POSIX Threads (Pthreads) and Windows Threads by Microsoft. While the syntax is different between the two APIs, most of the functionality in one model will be found in the other. Each model will create and join threads and feature synchronization objects to coordinate execution between threads and control the access to shared resources by multiple threads executing concurrently.

Pthreads has a thread container data type of `pthread_t`. This type is the handle by which the thread is referenced. In order to create a thread and associate that thread with a function for execution, the `pthread_create()` function is used. A *pthread_t* is returned through the parameter list. When one thread must wait on the termination of some other thread before proceeding with execution, a call to `pthread_join()` is called with the handle of the thread to be waited upon.

Windows threads use the ubiquitous kernel object `HANDLE` type for the handle of a thread. The `CreateThread()` function will return the `HANDLE` of a spawned thread. If the code will be using the C run-time library, it is recommended that the alternate `_beginthreadex()` function be used to create new threads. These two functions have exactly the same set of parameters, but the latter is safer to use with regard to initialization of thread resources and more reliable in the reclamation of allocated resources at thread termination.

Making one thread wait for another thread to terminate is accomplished by calling `WaitForSingleObject()`. To be more precise, since any kernel object in a program is referenced through a HANDLE, this function will block the calling thread until the HANDLE parameter is in the signaled state. If the HANDLE is a thread, the object will be signaled when the thread terminates. What it means for a HANDLE to be signaled is different for each type of kernel object. Windows also provides the `WaitForMultipleObjects()` function that can be used to wait until one or all of up to 64 HANDLEs are in the signaled state. Thus, with a single function call, a thread can join multiple threads.

The two synchronization objects most commonly used with Pthreads are the mutex (`pthread_mutex_t`) and the condition variable (`pthread_cond_t`). Instances of these objects must first be initialized before use within a program. Besides providing functions to do this initialization, defined constants are included in the Pthreads library that can be used for default static initialization when objects are declared. Mutex objects can be held by only one thread at a time. Threads request the privilege of holding a mutex by calling `pthread_lock()`. Other threads attempting to gain control of the mutex will be blocked until the thread that is holding the lock calls `pthread_unlock()`.

Condition variables are associated (through programming logic) with an arbitrary conditional expression and are used to signal threads when the status of the conditional under consideration may have changed. Threads block and wait on a condition variable to be signaled when calling `pthread_cond_wait()` on a given condition variable. At some point in the execution, when the status of the conditional may have changed, an executing thread calls `pthread_cond_signal()` on the associated condition variable to wake up a thread that has been blocked. The thread that receives the signal should first check the status of the conditional expression and either return to waiting on the condition variable (conditional is not met) or proceed with execution. Signals to condition variables do not persist. Thus, if there is no thread waiting on a condition variable when it is signaled, that signal is discarded.

Windows threads provide two basic mutual exclusion synchronization objects: the mutex and the critical section. A mutex is a kernel object accessed and managed through a HANDLE. The `CreateMutex()` function will initialize a mutex object. To lock the mutex, `WaitForSingleObject()` is called; when the mutex handle is in the signaled state, the mutex is available, and the wait function will return. If a thread holding a mutex is terminated unexpectedly, the mutex will be considered abandoned. The next thread to wait on the mutex will be able to lock the mutex and the return code from the wait function will indicate that the mutex had been abandoned. Once a thread is finished with the mutex, `ReleaseMutex()` unlocks the object for another thread to gain control. Windows mutexes, like other kernel objects, can be shared between different processes to create mutually exclusive access to shared resources.

Windows `CRITICAL_SECTION` objects functions like mutexes, but they are only accessible within the process in which they have been declared. Critical section objects are initialized with the `InitializeCriticalSection()` function before use. The programmer places `EnterCriticalSection()` and `LeaveCriticalSection()` calls around critical regions of code with a reference to an appropriate `CRITICAL_SECTION`. While critical section objects will be unavailable through termination of the program if the thread holding the object is terminated, the overhead of using this method of mutual exclusion will be considerably faster than use of a mutex object.

Windows events are used to send signals from one thread to another in order to coordinate execution. Events are kernel objects and are manipulated by use of a HANDLE. Threads use one of the wait functions to pause execution until the event is in the signaled state. The `CreateEvent()` function initializes an event and selects the type of event. There are two types of events: manual reset and auto-reset. The `SetEvent()` function will set either type of event to the signaled state. Any and all threads waiting on a manual reset event, once it has been signaled, will return from the wait function and proceed. Plus, any thread that calls a wait function on that event will be immediately released. No threads will be blocked waiting on the signaled event until a call to `ResetEvent()` has been issued. In the case of auto-reset events, only one thread waiting or the first thread to wait for the event to be signaled will return from the wait function and the event will be automatically reset to the non-signaled state. Unlike condition variables in Pthreads, the signals to Windows events will persist until either reset or the required number of threads have waited for the event and been released.

4.3.4.2 Intel® Threading Building Blocks

Intel® TBB is a C++ template-based library for loop-level parallelism that concentrates on defining tasks rather than explicit threads. The components to TBB include generic parallel algorithms, concurrent containers, low-level synchronization primitives, and a task scheduler. TBB has been published as both a commercial version and an open source version [16].

Programmers using TBB can parallelize the execution of loop iterations by treating chunks of iterations as tasks and allowing the TBB task scheduler to determine the task sizes, number of threads to use, assignment of tasks to those threads, and how those threads are scheduled for execution. The task scheduler will give precedence to tasks that have been most recently in a core with the idea of making best use of the cache that likely contains the task's data. The task scheduler utilizes a task-stealing mechanism to load balance the execution.

The `parallel_for` template is used to parallelize tasks that are contained within a for-loop. The template requires two parameters: a range type over which to iterate and a body type that iterates over the range or a subrange. The range class must define a copy constructor and a destructor, `is_empty()` (returns TRUE if the range is empty) and `is_divisible()` (returns TRUE if the range can be split) methods, and a splitting constructor (to divide the range in half). Besides the lower and upper bounds for the range, the range type also requires a grain size value to determine the smallest number of iterations that should be contained within a task. The TBB library contains two pre-defined range types: `blocked_range` and `blocked_range2D`. These ranges are used for single- and two-dimensional ranges respectively.

The body class must define a copy constructor and a destructor as well as the `operator()` method. The `operator()` method will contain a copy of the original serial loop that has been modified to run over a subrange of values that come from the range type.

The `parallel_reduce` template will iterate over a range and combine partial results computed by each task into a final (reduction) value. The range type for `parallel_reduce` has the same requirements as the `parallel_for`. The body type needs a splitting constructor and a `join()` method. The splitting constructor in the body is needed to copy read-only data required to run the loop body and to assign the identity element of the reduction operation that initializes the reduction variable. The `join()` method combines partial results of tasks based on the reduction operation being used. Multiple reductions can be computed simultaneously, for example, the minimum, maximum, and average of a data set.

Other generic parallel algorithms included in the TBB library are:

- `parallel_while`, to execute independent loop iterations with unknown or dynamically changing bounds

- `parallel_scan`, compute the parallel prefix of a data set; *pipeline*, for data-flow pipeline patterns

- `parallel_sort`, an iterative version of Quicksort that has been parallelized

Intel TBB also defines concurrent containers for hash tables, vectors, and queues. The C++ STL containers do not allow concurrent updates. In fact, between the point of testing for the presence of data within such a container and accessing that data it is possible that a thread can be interrupted; upon being restored to the CPU the condition of the container may have been changed by intervening threads in such a way that the pending data access will encounter undefined values and might corrupt the container. The TBB containers are designed for safe use with multiple threads attempting concurrent access to the containers. Not only can these containers be used in conjunction with the TBB parallel algorithms, but they can be used with native Windows or POSIX threads.

Mutex objects on which a thread can obtain a lock and enforce mutual exclusion on critical code regions are available within TBB. There are several different types of mutexes of which the most common is a `spin_mutex`. A `queuing_mutex` variation that is scalable (tends to take the same amount of time regardless of the number of threads) and fair – both properties that a `spin_mutex` does not have – is the other type available. There are also reader-writer lock versions of these two mutex types. None of these mutex types are reentrant. The other type of synchronization supported by TBB is atomic operations. Besides a small set of simple operators, there are atomic methods `fetch_and_store()` (update with given value and return original), `fetch_and_add()` (increment by given value and return original), and `compare_and_swap()` (if current value equals second value, update with first; always return original value).

A good resource covering TBB in more detail is from Reinders [17].

4.3.4.3 Multi-threaded Domain-Specific Libraries

Threaded libraries are becoming more prevalent to take advantage of the computing power available within multi-core processors. Two such libraries are the Intel® Math Kernel Library (MKL) and Intel® Integrated Performance Primitives (IPP) [18]. There are five distinct sections to the MKL: Basic Linear Algebra Subroutines (BLAS), Linear Algebra Package (LAPACK), Discrete Fourier Transforms, Vector Math Library (VML), and Vector Statistics Library (VSL). From the first two sections, the amount of computation versus thread management costs limits threading to the Level 3 BLAS routines along with select LAPACK and Fixed Fourier Transforms (FFT) routines. Other routines from the VML and VSL sections are also threaded depending on the routine and processor that will be used for execution. All threading within the MKL is done with OpenMP, and all routines within the library are designed and compiled for thread safety. The Cluster MKL version of the library includes all of the functions from the above sections plus the scalable LAPACK (ScaLAPACK) and Basic Linear Algebra Communication Subroutines (BLACS) functions.

The Intel IPP library contains a broad range of functionality. These areas include image processing, audio and video coding, data compression, cryptography, speech coding and recognition, and signal processing. Due to the number of functions within the IPP, different processing areas are supported by separate library linkable files. Dynamic libraries are threaded internally; static libraries are not threaded. Whether or not a library version is threaded, all functions within the IPP library are thread-safe. Further information on Intel IPP can be found in the book from Taylor [19].

4.3.4.4 C++ Thread Library

The C++ standards committee is pursuing the addition of a threading library for inclusion in the next C++ standard, currently termed C++0X. Several proposals are being investigated [20] and the ones under discussion provide explicit threading capabilities similar to Pthreads, but using and fitting in with inherent C++ characteristics such as classes, templates, and exceptions. Threads, mutexes, locks, conditions variables, and *thread pools* are under discussion.

4.3.5 Graphical Design Tools

Domain-specific graphical design tools such as National Instruments* LabVIEW* 8.5 allow users to visually diagram computations in an embedded application. LabVIEW offers an easy path to taking advantage of multi-core processors. The coding constructs enforce the program to be in what is termed structured dataflow where dependencies between inputs and outputs are known and pointer aliasing is prevented. These properties make it simple to visually determine what parts of a LabVIEW program can execute in parallel.

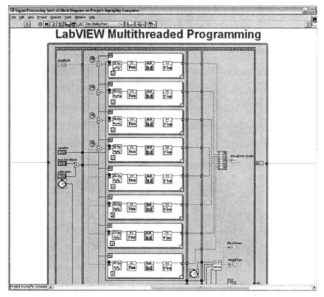

Figure 4.8: National Instruments LabVIEW

Figure 4.8 depicts a LabVIEW program to compute FFTs in parallel. Each of the eight rectangular boxes, termed a sub VI, computes an FFT and is independent of the other FFT calculations. Figure 4.9 depicts the running application on a system with two quad core processors. The application window (on the left) shows a speedup of 7.2X for the application as a result of the partitioning and subsequent execution on the eight processor cores. The application is embarrassingly parallel; however, it serves to illustrate the power of the graphical design tool in that the programmer does not explicitly worry about typical threading issues such as data races.

Instead the programmer communicates the flow of data into and out of various functions and the compilation and run-time system translates graphical dataflow design into an executable that enforces correct threading and efficient synchronization featuring:

- Underlying libraries that are both thread-safe and reentrant

- Device drivers designed for efficient performance

- OS supports load balancing, multi-threading, and multi-tasking

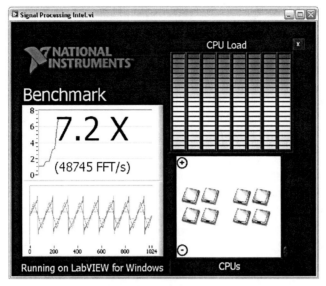

Figure 4.9: LabVIEW Virtual Instrument

Additional information on LabVIEW can be found in a whitepaper series [21] and a general topics paper [22].

4.3.6 Debugger

The next sections details tools to debug multi-threaded programs including traditional debuggers extended to handle threads and thread verification tools.

4.3.6.1 Multi-core Aware Debugging

Two popular Linux debuggers, dbx and gdb, are thread-aware and can be used to track down logic errors that are not related directly to the threaded implementation of the code. In dbx, the `thread` subcommand is used to display and control user threads. If no parameters are given to this command, the information of all user threads is displayed. Optionally, if thread numbers are given as parameters, information from the chosen threads is displayed. Thread execution can be held and released using `thread hold` and `thread unhold` respectively. Both subcommands apply to all threads if no parameters are given, or to the chosen threads with the given thread numbers. In order to examine the current status of a thread's execution with `print`, `registers`, and `where`, it must be set to the current thread by first issuing the command `thread current <threadnumber>`. A list of which threads are in the run, suspended, wait, or terminated state will be printed by using the `run`, `susp`, `wait` and `term` flags on the `thread` subcommand. The `mutex` and `condition` subcommands are used to display information on mutexes and condition variables.

The gdb debugger will notify the user when a new thread is spawned during the debug session. The `thread` command with a thread number parameter will set the chosen thread as the current thread. All commands requesting information on the program are executed within the framework of the current thread. Issuing an `info threads` command will display the current status of each thread within the program that shows the gdb-assigned thread number, the system's thread identifier, and the current stack frame of the thread. An asterisk to the left of the thread number indicates the current thread. In order to apply a command to more than the current thread, the `thread apply` command is used. This command takes either a single thread number, a range of thread numbers, or the keyword `all` before the command that should be applied to the designated threads. Breakpoints can be assigned to specific threads through the `break <linespec> thread <threadnumber>` command.

Other debuggers that are able to debug multi-threaded applications are the Intel Debugger (idb) and Totalview[*] from Totalview Tech. The Intel Debugger has dbx and gdb emulation modes that implement many of the thread-specific commands available in those debuggers. Totalview is able to debug multiple processes that are executing within a distributed, message-passing environment through MPI. Also, within a chosen process, individual threads can be selected, examined, and controlled through the GUI.

4.3.6.2 Thread Verification Tools

Data races are the most common cause of error in multi-threaded applications. They can also be the hardest to isolate because of the non-deterministic scheduling of thread execution by the OS. Running a threaded code on the same system during development and tests may not reveal any problems; yet, running the same binary on another system with any slight difference that could affect the order of thread execution may yield unexpected or erroneous results on the very first execution. The Intel® Thread Checker is a tool designed to identify data races, potential deadlocks, thread stalls, and other threading errors, within a code threaded with OpenMP, POSIX, or Windows threads.

As a plug-in to the Intel® VTune™ Performance Analyzer, Intel Thread Checker runs a dynamic analysis of a threaded application as it executes. To find data races, for example, the tool watches all memory accesses during threaded execution. By comparing the addresses accessed by different threads, and whether or not those accesses are protected by some form of synchronization, read–write and write–write conflicts can be found. Dynamic analysis will catch both obvious errors of accessing the variables visible to multiple threads, as well as memory locations accessed indirectly through pointers.

Of course, in order to watch memory accesses, Thread Checker must insert instrumentation within the application for that purpose. The instrumentation can be inserted directly into the binary file (*binary instrumentation*) just before the analysis is run or inserted at the time of compilation (source instrumentation) if using an Intel Compiler. Further information on the instrumentation and verification technology can be found in the paper by Banerjee et al. [23].

Two other tools with similar capabilities include Helgrind [24] and Thread Validator [25].

4.3.7 Performance Analysis Tools

This section summarizes traditional function profiling tools and thread profiling.

4.3.7.1 Profiling

The purpose of profiling the execution of an application is to find the hotspots of that application. A hotspot is simply a section of code that contains a large amount of activity. Typically this activity is time spent during the run of the application. However, activities such as cache misses, page faults, and context switches can be tracked with more sophisticated tools. The hotspots are indicators of where attention needs to be spent in order to optimize the code to reduce the impact of negative activities. If the user wants to thread serial code, time spent will be the activity of interest. Those parts of the code that take the most execution time (assuming that some other optimization would not reduce the time taken) are good candidates for threading, since these hotspots are going to be the most computationally intensive portions of the serial code.

A common Linux profiling tool is `gprof`. Actually, `gprof` is the display tool for data collected during the execution of an application compiled and instrumented for profiling. The -pg option, used in the `cc` command, will instrument C code. The instrumented binary will generate a profile data file (gmon.out is the default) when run. The *call graph* data that `gprof` prints includes the amount of time spent in the code of each function and the amount of time spent in the child functions called. By default, the functions are arranged in order of execution time, from largest to smallest. This order gives the user a ranked list of the functions that should be further examined for optimization or for parallelization by threads.

The Intel® VTune™ Performance Analyzer has two primary collectors: *event-based sampling* (EBS) and call graph. During EBS runs of the application, the collector periodically interrupts the processor when triggered after a number of microarchitectural events have occurred. Typically this will be ticks of the system clock, but can be set to be triggered by many different architectural events. During the interrupt, the collector records the execution context, including the current execution address in memory, OS process and thread ID executing, and executable module loaded at that address. Once execution of the target application has completed, the sampling data for the entire system (and the processes that were running during the sampling run) is displayed within the VTune Performance Analyzer GUI. Hotspot data can be found at the function and even source line level (given the proper compilation and link flags were used).

The call graph collector within the Intel VTune Performance Analyzer is similar to the `gprof` profiler. The target application is instrumented just before execution from within the VTune Performance Analyzer. Unlike the sampling collector, which can take samples from any process or module running during collection, only the application of interest will be profiled. The instrumentation records the caller of a function, how much time was spent within a function, and what functions were called, as well as the time spent within those child function calls. The function timing results of call graph are available within a table format, but can also be viewed as a graphical representation of the call tree structure resulting from the application run. Function execution time and number of callers or callees can be found by placing the mouse pointer over different parts of the displayed portions of the call tree. The call sequence that leads to the function with the longest execution time, known as the *critical path*, is highlighted by using red arcs to the function. This provides a graphic indication about what parts of the application should be considered for optimization or threading.

4.3.7.2 Thread Profiling

Besides viewing the collected sampling data as an aggregate over the course of the entire execution time that was sampled, the Intel VTune Performance Analyzer can display the sampling results over time. That is, the number of samples taken and associated with selected modules can be tallied within discrete time units during the sampling time. In this way, the load balance between threads of an application can be measured. If some threads have more samples taken during a given time range than some other threads, the latter threads will have typically done more computation within that time frame. While such load imbalances can be deduced from the aggregate data, the sample over time feature allows the user the opportunity to find the section(s) of code – down to source lines – that are the cause.

The Intel® Thread Profiler is a more general tool for identifying performance issues that are caused by the threading within an application. Intel Thread Profiler works on codes written with OpenMP, POSIX, or Windows threads. Within the OpenMP interface, aggregate data about time spent in serial or parallel regions is given via a histogram. Also, time spent accessing locks or within critical regions or with threads waiting at implicit barriers for other threads (imbalance) is also represented in the histogram. The summary information can be broken down to show execution profiles of individual parallel and serial regions, as well as how individual threads executed over the entire run. The former display is useful for finding regions that contain more of the undesired execution time (locks, synchronized, imbalance), while the latter is useful to discover if individual threads are responsible for undesired execution.

For an explicit threading model, Intel Thread Profiler employs *critical path analysis.* This is unrelated to the critical path of call graph analysis within VTune Performance Analyzer. As the application executes, the Intel Thread Profiler records how threads interact with other threads and notable events, such as spawning new threads, joining terminated threads, holding synchronization objects, waiting for synchronization objects to be released, and waiting for external events. An *execution flow* is the execution through an application by a thread and each of the listed events above can split or terminate the flow. The longest flow through the execution, the one that starts as the application is launched until the process terminates, is selected as the critical path. Thus, if any improvement in threaded performance along this path were done, the total execution time of the application would be reduced, increasing overall performance.

The data recorded along the critical path is the number of threads that are active (running or able to be run if additional core resources were available) and thread interactions over synchronization objects. Figure 4.10 depicts the Intel Thread Profiler GUI which

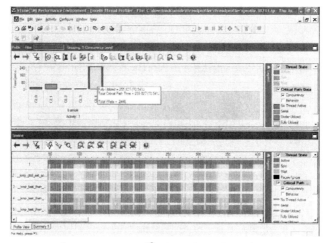

Figure 4.10: Intel® Thread Profiler views

contains two major divisions to display the information gathered during the threaded
execution: Profile View and Timeline View. The Profile View (top half of the screen)
gives a histogram representation of data taken from the critical path. This histogram can
be organized with different filters that include concurrency level (how many threads
were active along the critical path), object view, (what synchronization objects were
encountered by threads), and threads view (how each thread spent time on the critical
path). By use of the different filters and views the user can tell how much parallelism
was available during the application execution, locate load imbalances between threads,
and determine what synchronization objects were responsible for the most contention
between threads. The Timeline View (bottom half of the screen) shows the critical
path over the time that the application ran. The user can see the critical path switch
from one thread to another and how much time threads spent executing or waiting for a
synchronization object held by another thread.

Chapter Summary

This chapter served two important purposes. First, migration issues when moving your
embedded application to an x86 processor were discussed. Issues regarding OS selection,

32-bit versus 64-bit, and endianness were detailed. Unique considerations for embedded multi-core processors with regard to system level programming were also detailed. Many of these issues are independent of leveraging multi-core processors, but important nonetheless. Second, tools support for taking advantage of a multi-core x86 processor was discussed covering compiler support, thread libraries, multi-core debugging, and profiling. This high level knowledge is prerequisite to further chapters which employ these tools in the embedded multi-core case studies.

Related Reading

A good reference defining many wireless telecommunication terms is found in *Who Makes What: Wireless Infrastructure* [26].

References

[1] 64-bit Programming Models: Why LP64?, http://www.unix.org/version2/whatsnew/lp64_wp.html

[2] L. Matassa, *Endianness Whitepaper*, http://www.intel.com/design/intarch/papers/endian.pdf

[3] Transitive QuickTransit Technology, http://www.transitive.com/index.htm

[4] *AltiVec/SSE Migration Guide*, http://developer.apple.com/documentation/Performance/Conceptual/Accelerate_sse_migration/Accelerate_sse_migration.pdf

[5] Unified Extensible Firmware Interface, http://www.uefi.org/home

[6] Extensible Firmware Interface, http://www.intel.com/technology/efi

[7] Platform Support for MontaVista Linux, http://www.mvista.com/boards.php

[8] Support BSPs, http://msdn2.microsoft.com/en-us/embedded/aa714506.aspx

[9] Ardence, www.ardence.com/embedded/products.aspx?ID=70

[10] Board Support Packages, http://www.qnx.com/products/bsps/index.html

[11] *Intel® 64 and IA-32 Architectures Software Developer's Manual Volume 3A: System Programming Guide*, Part 1, http://www.intel.com/design/processor/manuals/253668.pdf

[12] AP-485 Intel® Processor Identification and the CPUID Instruction, http://developer.intel.com/design/xeon/applnots/241618.htm

[13] OpenMP, www.openmp.org

[14] J. Collins, H. Wang, D. Tullsen, C. Hughes, Y. Lee, D. Lavery, and J. Shen, *Speculative Precomputation: Long-range Prefetching of Delinquent Loads*, http://www.intel.com/research/mrl/library/148_collins_j.pdf

[15] Intel Threading Building Blocks 2.0, http://www.intel.com/cd/software/products/asmo-na/eng/threading/294797.htm

[16] Intel Threading Building Blocks 2.0 for Open Source, http://www.threadingbuildingblocks.org

[17] J. Reinders, *Intel Threading Building Blocks: Outfitting C++ for Multi-core Processor Parallelism*. Sebastopol: O'Reilly Media, Inc., 2007.

[18] Intel Integrated Performance Primitives, http://www.intel.com/cd/software/products/asmo-na/eng/perflib/ipp/302910.htm

[19] S. Taylor, *Intel® Integrated Performance Primitives – How to Optimize Software Applications Using Intel® IPP*. Santa Clara, CA: Intel Press, 2003.

[20] *Multi-threading Proposals from JTC1/SC22/WG21 –* Paper 2007, http://www.open-std.org/jtc1/sc22/wg21/docs/papers/2007/n2320.html, http://www.open-std.org/jtc1/sc22/wg21/docs/papers/2007/n2276.html, http://www.open-std.org/jtc1/sc22/wg21/docs/papers/2007/n2185.html

[21] Multicore Programming Fundamentals Whitepaper Series, http://zone.ni.com/devzone/cda/tut/p/id/6422

[22] National Instruments LabVIEW General Topic Overview, http://zone.ni.com/devzone/cda/pub/p/id/158

[23] U. Banerjee, B. Bliss, Z. Ma, and P. Petersen, A Theory of Data Race Detection, *International Symposium on Software Testing and Analysis, Proceeding of the 2006 workshop on Parallel and distributed systems: Testing and debugging*, Portland, Maine, http://support.intel.com/support/performancetools/sb/CS-026934.htm

[24] Helgrind, http://valgrind.org/info/tools.html#helgrind

[25] Thread Validator, http://www.softwareverify.com/cpp/thread/index.html

[26] *Who Makes What: Wireless Infrastructure*, http://www.lightreading.com/document.asp?doc_id=96166&print=true

## CHAPTER 5

Scalar Optimization and Usability

Key Points

- Before beginning a parallel optimization project, make sure a good effort at scalar performance optimization has already been invested.

- A compiler that targets your processor and features advanced optimizations such as automatic vectorization, interprocedural optimization, and profile-guided optimization can substantially improve performance of your application.

- Applying usability features of your compiler such as those aiding compatibility, compile time, and code size, can substantially improve development efficiency.

As discussed in Chapter 4, a variety of tools are available to help with multi-core software development. After reading that chapter you may be led to believe the next step is to start using the tools and begin developing parallel applications, but this is not the case. Before taking advantage of parallelism via multi-threading or partitioning, it is important to consider *scalar optimization* techniques. To say it more strongly, a prerequisite for parallel optimization is highly tuned scalar performance. Why is this claim true?

Consider a hypothetical performance optimization project with the following requirement:

Parallel optimization must provide a 30% increase in performance over the scalar version of the application.

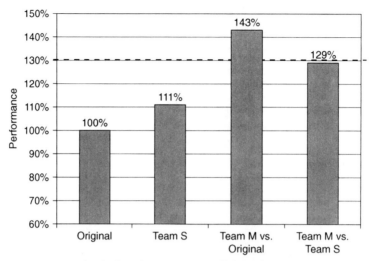

Figure 5.1: Hypothetical scalar versus parallel performance improvement

A development team, Team M, is created to develop a prototype of the application that employs parallel optimization and multi-core processors to meet the performance requirement. Another team, Team S, is created to see how much scalar optimization techniques can improve performance. Each team prototypes their improvement and a performance comparison is obtained. Figure 5.1 is a graphical representation of the performance obtained by the different teams. As you can see, Team M increased performance by 43% over the original code. Team S increased performance by 11% over the original code. The question – did Team M meet its goal?

The last column in the graph shows the performance improvement comparing Team M's results against Team S, which can be considered a new scalar version of the application.

Strictly speaking, Team M did not meet the goal. The performance difference between Team M and the new scalar version (Team S) is only 29%. Now it could be argued that the goal should have been clear in specifying that the original version of the code should be used for the performance comparison, but the reality is we are discussing end results. If the scalar optimization work could be accomplished with minimal effort and the parallel optimization effort required a large amount of resources, the parallel optimization effort may be terminated. If Team M had known about the performance headroom

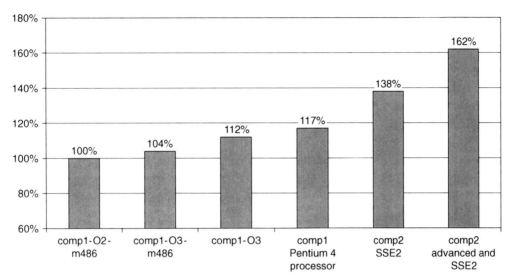

Figure 5.2: Compiler and target architecture comparison

offered by scalar optimization, Team M could have perhaps applied parallel techniques on more of the application to meet the performance goal. One key point is:

Before beginning a parallel optimization project, make sure a good effort at scalar performance optimization has already been invested.

One of the better and easier methods of increasing the scalar performance of your application is to apply aggressive compiler optimization.

C and C++ compilers are widely used for embedded software development[1]. Compiler optimization can play a large role in increasing application performance and employing the latest compiler technology specifically targeting your processor can provide even greater benefit. Figure 5.2 is a comparison of two different compilers and several optimization settings executing SPEC CINT2000 [1] on an Intel® Pentium® M processor system. The two compilers employed are labeled "comp1" and "comp2" respectively[2]. Four different optimization settings were employed when comp1 compiled the benchmark. Two different optimization settings were used when comp2 compiled the

---

[1] Currently C is widely used. C++ is growing in usage.
[2] The specific versions of the compilers are not relevant to the discussion.

benchmark. The compiler, comp1, targeting an Intel486 processor serves as the baseline and is represented by the far left bar in the graph. Using comp1 with the −O3 option and targeting an Intel486 processor produces a 4% performance improvement. The −O3 option provides greater optimization over the default optimization setting. Using comp1 with the −O3 option and its default *compiler processor target* results in a 12% improvement. Using comp1 with the processor target of an Intel® Pentium® 4 processor results in a 17% performance improvement. Employing the second compiler, comp2, and targeting the Pentium M processor leads to a 38% performance improvement. Finally, using comp2 targeting the Pentium M processor and *advanced optimization* leads to a 62% improvement. The baseline compiler and option setting (comp1 targeting an Intel486 processor) represents legacy applications that employ dated compilation tools and target older processors while relying solely upon new processors to provide application performance improvement. The comp2 targeting the Pentium M processor and using advanced optimization represents a compiler with the latest optimization technology and optimization that specifically targets the processor that is used to execute the application in the field. Terms such as advanced optimization and processor target are explained more fully in latter sections of this chapter. Two points should be clear from this data:

- Legacy applications that have not been recompiled for new processor targets may be sacrificing performance.

- Employing a high-performance compiler specifically tuned for your architecture can lead to big performance improvements.

For developers that use C and C++, knowing the performance features available in their respective compiler is essential. This chapter describes performance features available in many C and C++ compilers, the benefits of these features, and how to employ them. The first section details C and C++ compiler performance features such as general optimizations, advanced optimizations, and user-directed optimization. The second section details a process to use when optimizing your application. The third section of the chapter is a case study that employs the optimization process to optimize a sample application. The final section discusses usability features that can aid compatibility, and methods for reducing compile time and code size. With the techniques described in this chapter, embedded developers can extract higher performance from their applications with less effort.

5.1 Compiler Optimizations

A compiler optimizes by analyzing source code and determining a representation of the source code in machine language that executes as effectively as possible. Compiler optimizations can be grouped into general optimizations and advanced optimizations:

- General optimizations include both architecture-independent and architecture-dependent optimizations.

- Advanced optimizations are specialized and show the highest benefits on very specific types of code.

5.1.1 General Optimizations

General optimizations are comprised of architecture-independent and architecture-dependent optimizations. Architecture-independent optimizations do not rely upon knowledge of the underlying architecture; these are good techniques to apply no matter what platform the code will execute on. Examples of architecture-independent optimizations include dead code elimination, common-subexpression elimination, and loop-invariant code motion. Figure 5.3 shows code where the compiler may perform the following optimizations to yield higher performance code:

- Dead Code Elimination – lines 6 and 7 could be optimized away, because the if statement at line 6 can be proved to always be false guaranteeing that line 7 can never execute.

- Loop-invariant code motion – the compare at line 9 could be moved outside of the loop, because the value of *y* is not affected by anything else in the loop.

- Common-subexpression elimination – the pointer calculations for the two references of a[i] at Line 10 could be shared.

Architecture-dependent optimizations include *register allocation* and *instruction scheduling*. A detailed knowledge of the microprocessor architecture is required to create a good architecture-dependent optimizer. Register allocation is compiler functionality that determines where variables are loaded for computation in the processor. The x86 ISA has only eight general purpose registers so the number of active computations that can be in progress at the same time is somewhat limited. Instruction scheduling is the ordering of machine language code based upon internal processor constraints and

```
Line                          Code
   1  #define DEBUG 0
   2  int a[100];
   3  int foobar(int y)
   4  {
   5    int x=0;
   6    if (DEBUG)
   7       printf("In foobar\n");
   8    for (i=0;i<100;i++) {
   9       if (y==0)
  10           x += a[i] + (2 * a[i]);
  11    }
```

Figure 5.3: General optimization code sample

Table 5.1: Description of general optimization options

Optimization	Description
-O2	Use good baseline architecture-independent optimizations.
-O3	Use more aggressive architecture-independent optimizations over -O2 that may trade-off compile time or code size for executable speed.
-mtune=	Schedule low level instructions that are tuned to the instruction latencies of a given processor. The code will still execute on other older members of the processor family.
-march=	Optimize and generate code that uses instructions that may not be available on older members of the processor family.

program dependencies. Both register allocation and instruction scheduling is the domain of compiler writers and assembly language programmers.

Typically, combinations of general optimizations are bundled under a few compiler options. Table 5.1 summarizes a variety of general optimizations available in a number of compilers that can be used to optimize your application. The option -O2 is a good basic optimization flag and is the default optimization setting in some compilers. The -O3 option employs more aggressive optimizations that can improve performance, but may cost more in terms of application code size or compilation time. Consider these general optimization options as a first step in performance tuning of your application.

The -mtune= option schedules instructions based upon the processor specified. For example, to schedule instructions for a Pentium M processor, you would use -mtune=pentium-m. This option may result in lower performance on architectures that differ from the one specified. The -march= option generates code with a specific processor in mind. This option automatically sets -mtune= to the same processor and in addition may use instructions that are not supported on older processors. For example, the -mtune=prescott option may select SSE3 instructions which would not execute on a processor that does not support SSE3 such as the Pentium III processor. Consider the -mtune= option if the majority of your target systems are a specific processor target. Consider the -march= option if all of your target systems are feature equivalent with the processor specified.

5.1.2 Advanced Optimizations

Some optimizations take advantage of more recent technologies in a processor and/or require more effort on the part of the developer to use. The term advanced optimizations is used to collectively refer to these compiler features. The next sections describe three such advanced optimizations: *automatic vectorization, interprocedural optimization*, and *profile-guided optimization*.

5.1.2.1 Automatic Vectorization

Several processors have extended their instruction set to allow access to new capabilities in the processors such as data prefetching and parallel execution. For example, the Embedded Intel Architecture processors offer a set of instruction extensions termed MMX™ technology, SSE, SSE2, SSE3, SSSE3, and SSE4. For higher-level languages such as C and C++, compiler technology provides the gateway to these new instructions. Methods of employing these instructions in your application may include:

- Writing inline assembly language to explicitly specify the new instructions

- Employing a C intrinsic library or a C++ class library, giving access to higher-level language instructions

- Employing automatic vectorization technology to automatically generate the new instructions

```
void vector_mul(double * restrict a, double * restrict b,
double * restrict c){
  int i;
  for (i=0;i< 100;i++) {
    a[i] = b[i] * c[i];
  }
  return;
}

..B1.6:                                    # Preds ..B1.4 ..B1.6
        movsd     (%ebp,%edx,8), %xmm0
        movhpd    8(%ebp,%edx,8), %xmm0
        mulpd     (%ecx,%edx,8), %xmm0
        movapd    %xmm0, (%esi,%edx,8)
        addl      $2, %edx
        cmpl      %eax, %edx
        jb        ..B1.6
```

Figure 5.4: Vectorization example – C source and resulting assembly code

Automatic vectorization technology enables the compiler to analyze your C and C++ source code to determine where these instructions can be employed to speed up your code. A compiler that performs vectorization analyzes loops and determines when it is safe to execute several iterations of the loop in parallel.

Figure 5.4 is a C source code example and the resulting x86 assembly language when compiling the code with the Intel® C++ compiler using automatic vectorization. The vectorizer is able to take advantage of SSE2 instructions and transforms the loop to compute two results of a[i] per iteration, instead of one result as specified by the source code.

5.1.2.2 Interprocedural Optimization

Compilers typically process one function at a time and in isolation from other functions in the program. During optimization, the compiler must often make conservative assumptions regarding values in the program due to side-effects that may occur in other functions, limiting the opportunity for optimization. Use a compiler with interprocedural optimization to optimize each function with detailed knowledge of other functions in

```
1       int check (int debug) {
2         for (int i=0;i<100;i++) {
3           if (debug)
4               return 1;
5         }
6         return 0;
7       }
8
9       int main() {
10        check(0);
11      }
```

Figure 5.5: Interprocedural optimization code sample

the application. Interprocedural optimization enables other optimizations; these other optimizations are more aggressive because of the enhanced interprocedural information. Examples of typical optimizations enabled by interprocedural optimization are:

- Inlining – the ability to inline functions without any user direction.

- Arguments in registers – function arguments can be passed in registers instead of the stack which can reduce call/return overhead.

- Interprocedural constant propagation – propagate constant arguments through function calls.

- Loop-invariant code motion – better compiler analysis information can increase the amount of code that can be safely moved outside of loop bodies.

- Dead code elimination – better global information can increase detection of code that can be proved to be unreachable.

Figure 5.5 is an example of code that can be optimized effectively with interprocedural optimization. In the example, the entire body of the loop in the function check() can be optimized away for the following reasons:

- Interprocedural constant propagation will propagate the argument to the called function at line 10 which is constant in the function check().

- Loop-invariant code motion will recognize that the if statement at line 3 is not dependent on the loop.

- Dead code elimination will eliminate the code under the true condition of the if statement at line 3 because the variable debug is a constant, zero.

5.1.2.3 Profile-guided Optimization

Profile-guided optimization enables the compiler to learn from experience. Profile-guided optimization is a three-stage process:

- First compile the software for profile generation. The compiler inserts instrumentation into the compiled code able to record metrics about the behavior of the code when it is executed.

- Execute the application to collect a profile that contains characteristics of what is termed a "training run" of the application.

- Recompile the application to take advantage of the profile or what has been learned about the application during the training run.

The compiler now knows where the most frequently executed paths in the program reside and can prioritize its optimization to those areas. A description of several optimizations enabled by profile-guided optimization in a typical compiler follows:

- Function ordering – improves instruction cache hit rates by placing frequently executed functions together.

- Switch-statement optimization – optimizes the most frequently executed cases in a switch statement to occur first.

- Basic block ordering – improves instruction cache hit rates by placing frequently executed blocks together.

- Improved register allocation – gives the best register assignments to calculations in the most highly executed regions of code.

Chapter 2 included an example of switch-statement optimization where the determination of the frequently executed cases of a switch statement led to improved layout of the machine language code.

Table 5.2: Advanced optimization options

Compiler	Automatic vectorization	Interprocedural optimization	Profile-guided optimization
GNU gcc	`-ftree-vectorize`	`-whole-program` `--combine`	`-fprofile-generate` `-fprofile-use`

Microsoft Visual C++	Not available	`/GL`
		`/LTCG`
			`/GL`
			`/LTCG:PGI`
			`/LTCG:PGO`

Intel C++ compiler for Linux	`-x[T,P,O,N,W,K]`	`-ipo`	`-prof-gen` `-prof-use`

5.1.3 Advanced Optimization Options

The options to enable advanced optimization in several compilers are listed in Table 5.2.

The GNU gcc version 4.1 features advanced optimizations. Automatic vectorization is enabled with the `-ftree-vectorize` option. General interprocedural optimization is not available; however, some interprocedural optimization is enabled with the `-whole-program` and `--combine` options. Profile-guided optimization is enabled with the `-fprofile-generate` option for profile generation and the `-profile-use` option for the profile use phase.

The Microsoft Visual C++ compiler does not currently support automatic vectorization. Interprocedural optimization is enabled with the /GL option for compilation and the /LTCG option at link time. Profile-guided optimization is enabled with /GL option for compilation, and linked with /LTCG:PGI for profile generation and /LTCG:PGO for profile application.

The Intel C++ Compiler for Linux enables automatic vectorization with the `-xproc` option where `proc` specifies one of several processor targets. Interprocedural optimization is enabled with the `-ipo` option on both the compile command line and link command line. Profile-guided optimization is enabled with the `-prof-gen` option for profile generation and `-prof-use` option for profile application.

The options listed above for each of the compilers are not precisely equivalent between them. For example, the gcc option `-ftree-vectorize` only enables automatic vectorization; however, the Intel C++ Compiler option -xproc enables automatic

```
int check (float *a, float *b, float *c) {
  for (int i=0;i<100;i+=1) {
    *a = *b + *c;
    a += 1; b += 1; c += 1;
  }
  return 1;
}
```

Figure 5.6: Aliasing code sample

vectorization and targets the specified processor so other optimizations such as software prefetch and instruction scheduling for the processor target may be enabled.

Read your compiler documentation to understand what is provided with your compiler advanced optimizations and any special requirements. Lastly, in comparing advanced optimization features of each compiler keep in mind the end value is application performance, which means that it may be necessary to build and measure the performance gained from each compiler and its advanced optimizations.

5.1.4 Aiding Optimizations

Developers can aid compiler optimization in situations where the compiler is restricted in terms of making assumptions about the code, how the code is laid out in memory, and when the developer has prior knowledge about the run-time behavior of the code. In these cases, careful understanding of these issues and how to effectively communicate optimization information to the compiler can have a substantial impact on performance.

The first area where developers can assist the compiler involves the use of pointers and aliasing issues that can restrict optimization. Aliasing occurs when two pointers reference the same object. The compiler must be conservative in its assumptions and optimizations in the presence of aliasing. Figure 5.6 shows a situation where the compiler cannot assume that *a* and *b* point at different regions of memory and must be conservative in optimizing or add run-time bound checks of the pointers. A compiler may even choose not to vectorize in this case.

The problem highlighted in Figure 5.6 can be solved by providing the compiler more information regarding the pointers referenced in the function. Several techniques are available for assisting the compiler in its *alias analysis* and are summarized as follows:

- Restrict – C99 defines the restrict keyword which allows the developer to specify pointers that do not point to the same regions of memory.

```
int check (float * restrict a,  float * restrict b,
           float *c) {
#pragma vector aligned
   for (int i=0;i<100;i+=1) {
      *a = *b + *c;
      a += 1; b += 1; c += 1;
   }
   return 1;
}
```

Figure 5.7: Restrict code sample

- Array notation – using array notation such as a[i] and b[i] helps denote referenced regions of memory.

- Interprocedural optimization – enables greater alias analysis and may enable the compiler to prove that certain pointers do not point to the same region.

The code in Figure 5.7 is based upon the code in Figure 5.6 and adds the use of the restrict keyword to allow the compiler to disambiguate *a* and *b* and to use a vectorized version of the loop at run-time.

Figure 5.7 also includes the use of a pragma, #pragma vector aligned, which communicates data layout information to the compiler. Some architectures contain instructions that execute faster if the data is guaranteed to be aligned on specific memory boundaries. Proper use of pragmas in communicating knowledge to the compiler can lead to several benefits as summarized in Table 5.3. For proper syntax and usage, refer to your compiler's reference manual.

One last area where developers can aid the compiler is with the memory layout of data structures used by the application. Modern architectures feature a memory hierarchy containing multiple levels of cache. Effective use of the memory hierarchy involves trying to keep data that is accessed very close together in time and space in contiguous regions. Techniques for optimizing for cache are numerous; however, a few of the important ones are summarized as:

- Data layout

- Data alignment

- Prefetching

Table 5.3: Typical compiler directives, attributes, and benefits

Pragma	Description and benefits
`#pragma vector {aligned, unaligned, always}`	Instructs the compiler to vectorize a loop. Increased performance on loops where the developer knows behavior. Can also communicate alignment information.
`#pragma novector`	Instructs the compiler to not vectorize a loop. Can result in code size savings. Performance savings on low-trip count loops.
`#pragma unroll`	Instructs the compiler to unroll a loop. Can result in increased performance, but larger code size.
`__declspec align (n)`	Instructs the compiler to align a variable on an n-byte boundary.
`__assume_aligned(a,n)`	Tells the compiler that *a* is aligned on an n-byte boundary.

```
struct {                          struct {
   char f;                           float a[SIZE];
   float a[SIZE];                    float b[SIZE];
   char z;                           float c[SIZE];
   float b[SIZE];                    char f;
   char v;                           char z;
   float c[SIZE];                    char v;
} soa;                            } soa2;
```

Figure 5.8: Data layout and alignment example

Data layout helps data structures fit in cache more effectively. Consider the structure declarations in Figure 5.8.

Suppose the value *SIZE* was defined to be 1. The size of *soa* on many compilers is 24 bytes because a compiler would pad three bytes after *f*, *z*, and *v*. The size of *soa2* on many compilers is 16 bytes which is a 33% reduction in size. Order the declarations of data in your structures to minimize unnecessary padding.

Data alignment is a technique that employs useful padding to allow efficient access to data. Consider again the declarations in Figure 5.8 where this time *SIZE* is defined to a large number, say 100,000. There is no guarantee that the arrays *a*, *b*, and *c* are aligned on a 16-byte boundary which prevents faster versions of SSE instructions to be employed if a loop that accesses the data is vectorized. For example, if the for loop detailed in

```
int foobar()
{
  int i;
  for (i=0;i<SIZE;i++) {
    soa.a[i] = soa.b[i] + soa.c[i];
  }
  return 0;
}
```

Figure 5.9: Unaligned load code sample

Figure 5.9 is vectorized using SSE instructions, unaligned loads will be performed on the data which are less efficient than their aligned counterparts. If you substitute *soa2* for *soa* in the code aligned loads will be employed.

Prefetch is a second technique to optimize memory access. The goal of prefetch is to request data to be placed in cache right before it is referenced by the application. Modern processors feature both software instructions to prefetch data and automatic prefetch. The automatic prefetch is a processor feature that analyzes the stream of memory references, attempts to predict future access, and places predicted memory accesses in the cache. Automatic prefetch helps in applications that have regular strided access to data and may hurt performance if the data access is irregular. In some cases, it may be beneficial to turn off automatic prefetch on your processor. Several BIOS programs enable the setting of automatic prefetch. If your application has irregular non-strided access to data, it may be worth turning off automatic prefetching to see if it increases performance.

5.2 Optimization Process

The previous section provided an overview of compiler optimization. This section details a process for applying these optimizations to your application. Figure 5.10 depicts the process which is comprised of 4 steps:

1. Characterize the application.

2. Prioritize compiler optimization.

3. Select benchmark.

4. Evaluate performance of compiler optimizations.

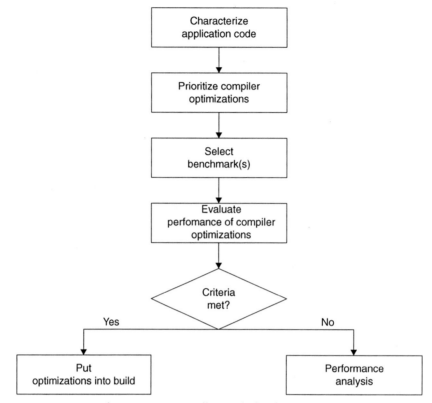

Figure 5.10: Compiler optimization process

Optimization using the compiler begins with a characterization of the application. The goal of this step is to determine properties of the code that may favor using one optimization over another and to help in prioritizing the optimizations to try. If the application is large it may benefit from optimizations for cache memory. If the application contains floating point calculations, automatic vectorization may provide a benefit. Table 5.4 summarizes the questions to consider and sample conclusions to draw based upon the answers.

The second step is to prioritize testing of compiler optimization settings based upon an understanding of which optimizations are likely to provide a beneficial performance increase. Performance runs take time and effort so it is essential to prioritize the optimizations that are likely to increase performance and foresee any potential challenges

Table 5.4: Application characterization

Question	If the answer is true
Is the application large or does the working data set exceed the size of the cache?	The application may be sensitive to cache optimizations.
Are there large amounts of numerical or floating point calculation?	Vectorization may help performance.
Does the source code make heavy use of classes, methods, and/or templates?	C++ code usually benefits from inlining.
Is the execution spread throughout the code in many small sections?	Code size optimizations may be beneficial.
Is the data access random, localized, or streaming?	Data access optimizations may be beneficial.

in applying them. For example, some advanced optimizations require changes to the build environment. If you want to measure the performance of these advanced optimizations, you must be willing to invest the time to make these changes. At the least, the effort required may lower the priority. Another example is the effect of higher optimization on debug information. Generally, higher optimization decreases the quality of debug information. So besides measuring performance during your evaluation, you should consider the effects on other software development requirements. If the debugging information degraded to an unacceptable level, you may decide against using the advanced optimization or you may investigate compiler options that can improve debug information.

The third step, select a benchmark, involves choosing a small input set for your application so that the performance of the application compiled with different optimization settings can be compared. In selecting a benchmark the following things should be kept in mind:

- The benchmark runs should be reproducible, for example, not result in substantially different times every run.

- The benchmark should run in a short time to enable running many performance experiments; however, the execution time cannot be so short that variations in run-time using the same optimizations are significant.

- The benchmark should be representative of what your customers typically run.

The next step is to build the application using the desired optimizations, run the tests, and evaluate the performance. The tests should be run at least three times apiece. My recommendation is to discard the slowest and fastest time and use the middle time as representative. I also recommend checking your results as you obtain them, seeing if the actual results match up with your expectations. If the time to do a performance run is significant, you may be able to analyze and verify your collected runs elsewhere and catch any mistakes or missed assumptions early. Finally, if the measured performance meets your performance targets, it is time to place the build changes into production. If the performance does not meet your target, the use of a performance analysis tool such as the Intel® VTune™ Performance Analyzer should be considered.

One key point in reading the proceeding sections: let the compiler do the work for you. There are many books that show you how to perform optimizations by hand such as unrolling loops in the source code. Compiler technology has reached a point now where in most cases it can determine when it is beneficial to perform loop unrolling. In cases where the developer has knowledge the compiler cannot ascertain about a particular piece of code, there are oftentimes directives or pragmas where you can provide the compiler the missing piece of information.

Case Study: Database

MySQL* [2] is an open source database widely used by enterprises to manage corporate data, handle transactions, and run e-commerce and data warehousing applications. MySQL is a key component of LAMP[3], a common web server solution consisting of Linux, Apache, MySQL, and PHP/Python/Perl. MySQL is available on many platforms including 32 bit and 64 bit, Linux, Unix, BSD, Apple, Windows, IA-32 Architecture, PowerPC, and SPARC. This case study focuses on MySQL version 4.1.12 running on Linux and a Pentium 4 processor-based system. MySQL running as part of LAMP and executing on a dedicated web server can be considered an embedded application in that the system is typically fixed function. This optimization process can be applied in a broader fashion targeting different applications and different processors.

---

[3] LAMP – Linux, Apache, MySQL, PHP.

Characterize the Application

MySQL is a database application; this fact suggests a number of easy characterizations. First, we expect the application to be large and it is – MySQL has several hundreds of thousand lines of code and an example combined code size for all applications in the client directory using the Intel Compiler on x86 is over 15 MB. Optimizations that aid code size may provide a benefit. Second, the application contains C++ source code so we will make sure inlining is used. Another question to ask is how much of the C++ language does the application use? MySQL uses classes but does not take advantage of C++ exception handling or run-time type identification. Options that can limit the amount of C++ feature overhead used should be considered. Third, databases typically access large amounts of memory and thus optimizations for data access may be beneficial. Finally, databases typically involve large amounts of integer-based calculations and little floating point calculations so optimizations geared to floating point such as automatic vectorization would not be expected to provide a performance increase.

Evaluate Compiler Optimization

The first optimization to try is the baseline optimization using -O2. For the Intel Compiler, the -O2 option enables a broad range of compiler optimizations such as partial redundancy elimination, strength reduction, Boolean propagation, graph-coloring register allocation, and sophisticated instruction selection and scheduling. In addition, single file inlining occurs at -O2 so we expect some of its benefits. Since inlining is important to C++ performance, we will also attempt more aggressive inlining by using single file interprocedural optimization (-ip) and multiple file interprocedural optimization (-ipo). The -ip option enables similar inlining to what is enabled at -O2, but performs a few more

Table 5.5: Intel compiler optimization evaluation

Optimization	Expectation
-O2	Baseline optimization
-O3	Data access optimizations should provide benefit
Single file interprocedural optimization (-ip)	Stronger inlining analysis over -O2
Multiple file interprocedural optimization (-ipo)	Multiple file inlining should bring further benefit
Profile-guided optimization (-prof-use)	Help performance through code size optimizations
Automatic vectorization (-xN)	No improvement is expected

analyses that should result in better performance. The `-ipo` option enables inlining across multiple files. Interestingly, inlining tends to increase code size; however, if the inlining results in smaller code size for the active part of the application by reducing call and return instructions, the net result is a performance gain. Profile-guided optimization (`-prof-use`) is a great optimization to use with inlining because it provides the number of times various functions are called and therefore guides the inlining optimization to only inline frequently executed functions, which helps reduce the code size impact. The `-O3` option enables higher-level optimizations focused on data access. The `-O3` option will be assessed to see what kind of performance benefits occur. Finally, it is expected that automatic vectorization would not provide a performance benefit; however, automatic vectorization is fairly easy to use so it will be attempted. The list of optimizations that will be attempted and the reasons for doing so are summarized in Table 5.5.

Select a Benchmark

A special benchmark called SetQuery [3] was developed to help in this optimization effort and was used to measure the performance of the MySQL database on Intel Architecture. SetQuery returns the time that the MySQL database takes to execute a set of SetQuery runs. The SetQuery benchmark measures database performance in a decision-support context such as data mining or management reporting. The benchmark calculates database performance in situations where querying the data is a key to the application performance as opposed to reading and writing records back into the databases.

Run Tests

The use of most of the optimizations was fairly straightforward; however, there are a few optimizations that require some effort to test. Multiple file interprocedural optimization essentially delays optimization until link time so that every file and function is visible during the process. To properly use `-ipo`, ensure the compiler flags match the link flags and that the proper linker and archiver are used. The original build environment defines the linker to `ld` and archiver to `ar` and these defines were changed to `xild` and `xiar` respectively.

One challenge in using profile-guided optimization is determining the correctness and proper use of the profile information by the compiler. The results were surprising and the obvious first suspicion is to question the sanity of the profiling data. Two techniques for verifying profile information are manually inspecting the compilation output and using the `profmerge` facility to dump profile information. During compilation, if the compiler does

Table 5.6: Sample profile-guided optimization output

Block execution count statistics				
cnt	/funcmaxent	/progmaxent	/progmaxblk	blk#
1	1	1	0.001	0
1	1	1	0.001	1
1000	1000	1000	1	2
1	1	1	0.001	3
999	999	999	0.999	4
1	1	1	0.001	5
0	.	.	.	6
1	1	1	0.001	7

not find profile information for some number of functions in the file that is being compiled, the compiler will emit the following diagnostic:

```
WARNING: field.cc, total routines: 667, routines w/profile
info: 9
```

The diagnostic in this example states that the file field.cc has a total of 667 routines and only 9 routines were executed during the profile generation phase. If the compiler is able to find profile information for all functions in a file, the compiler does not emit a diagnostic. Make sure the compiler either does not emit the above diagnostic or emits the diagnostic with a number of the functions in a file using profile information. If you are aware of the most frequently executed functions in your application and the file with those functions shows little to no routines compiled with profile information, the profile information may not be applied correctly. The second technique to verify profile information is to use the profmerge application with the -dump option as in:

```
profmerge -dump
```

which dumps the contents of the profile data file (pgopti.dpi). Search for a routine that is known to execute frequently and find the number of blocks in the function (BLOCKS:) and then the section "Block Execution Count Statistics" which contains counts of the number of times the blocks in the function were executed. You should expect to see a large number. Table 5.6 is an example output from the profile of a function that contains a loop that iterates 1000 times. Block 2 has a count of 1000 which lends confidence to the profile that was generated.

Table 5.7: Intel compiler optimization and SetQuery performance

Optimization	Code size (in bytes)	Code size increase (versus baseline) (%)	Execution time (in seconds)	Execution time improvement (versus baseline) (%)
-O2	16488449	0.00	526	0.00
-O3	16488449	0.00	520	1.15
-O3 -ip	16710369	1.35	484	8.68
-O3 -ipo	16729709	1.46	492	6.91
-O3 -ip -prof-use	16494273	0.04	544	-3.31
-O3 -ip -xN	16871105	2.32	487	8.01

Results

The performance runs using the Intel Compiler on x86 systems running Linux confirmed a number of our expectations; however, a few results surprised us. The use of stronger inlining with -ip resulted in higher performance over the baseline (8.68%) and a larger code size (1.35%). Surprisingly, the use of -ipo did not result in as great a performance improvement (6.91%) as -ip, but did result in larger code size (1.46%). The biggest surprise was the combination of -ip and -prof-use, which resulted in less performance than the baseline (−3.31%). True to expectation, the combination of -ip and -prof-use resulted in a code size increase of 0.04% which is better than the code size increase using -ip alone. Table 5.7 summarizes the results of several different option sets on a Pentium® 4 desktop system. The best performance was obtained by using -O3 -ip so we chose to use these two options in our *production build*. If greater performance is desired, analysis of why -ipo and -prof-use did not increase performance would be the first priority.

In addition, some of the aggressive optimization settings did not result in significant performance increases and in one case resulted in lower performance than the baseline optimization. These numbers are shared for a very important reason: stronger compiler optimizations do not guarantee better performance in all cases. This stresses the need to understand some level of detail in applying optimization to your application.

The open source database MySQL was optimized by using a set of compiler optimizations above and beyond the default -O2 optimization. By applying the process of characterizing the application, prioritizing compiler-optimization experiments, selecting a representative benchmark, and measuring performance results, it is possible to improve the performance of applications by taking advantage of higher levels of compiler optimization. This improvement does not require going to the depths of traditional performance analysis, however yields performance benefits nonetheless. Relying on aggressive compiler optimization is a practical first-cut technique for improving the performance of your application; if further performance gains are desired, the next step of low-level performance analysis should be considered.

5.3 Usability

Another area to improve before beginning an effort at parallelism is C and C++ compiler usability. The performance of your development environment and specifically your C and C++ compiler translates into increased developer performance due to the following:

- Greater standards compliance and diagnostics lead to fewer programming errors.

- Faster compilation times leads to more builds per day.

- Smaller application code size can enable more functionality to be provided by the application.

- Code coverage tools help build a robust test suite.

Performance of the generated code is a key attribute of a compiler; however, there are other attributes collectively categorized as usability attributes that play a key role in determining the ease of use and ultimately the success of development. The next several sections describe techniques for increasing the use of the compiler in terms of diagnostics, compatibility, compile time, code size, and code coverage. When employed correctly, these techniques can lead to increased developer efficiency and associated cost savings in the development phase of your embedded project.

5.3.1 Diagnostics

Two widely used languages for embedded software development are C and C++. Each of these languages has been standardized by various organizations. For example, ISO C conformance is the degree to which a compiler adheres to the ISO C standard. Currently, two ISO C standards are commonplace in the market: C89 [4] and C99 [5]. C89 is

Table 5.8: Description of diagnostic options

Optimization	Description
`-Wall`	Enable all warnings.
`-Werror`	Promote all warnings to errors.
`-Wcheck`	Enable a set of diagnostics that help diagnose issues that lead to hard to debug run-time issues.
`-we`	Promote a warning to an error.
`-ansi`	Check the code for ISO conformance.

currently the *de facto* standard for C applications; however, use of C99 is increasing. The ISO C++ standard [6] was approved in 1998. The C++ language is derived from C89 and is largely compatible with it; differences between the languages are summarized in Appendix C of the ISO C++ standard. Very few C++ compilers offer full 100% conformance to the C++ standard due to a number of somewhat esoteric features such as exception specifications and the export template. Most compilers provide an option to specify a stricter degree of ISO conformance. Employing an option to enforce ISO conformance helps developers who may be interested in porting their application to be compiled in multiple compiler environments.

Compiler diagnostics also aid porting and development. Many compilers feature an option to diagnose potential problems in your code that may lead to difficult to debug run-time issues. Use this option during the development of your code to reduce your debug time. One other useful diagnostic feature is the discretionary diagnostic. A discretionary diagnostic has a number associated with it which allows the promotion or demotion of the diagnostic between a compile-stopping error message, warning message, remark, or no diagnostic. Table 5.8 describes the options available in the Intel® C++ Compiler for Linux* that provides the abilities mentioned above.

Some compilers include a configuration file that allows the placement of additional options to be used during compilation and can be a useful location for additional discretionary diagnostics. For example, to promote warning #1011 to an error for all C compilations, placing `-we1011` in the configuration file eliminates the need to manually add the option to every compilation command line.

For full information on diagnostic features including availability and syntax, please see your compiler's reference manual.

5.3.2 Compatibility

One challenge in embedded software development is porting software designed and optimized for one architecture onto a new upgraded architecture within the same family. An example is upgrading the system that runs your application from an Intel® Pentium® III processor-based system to an Intel Pentium 4 processor-based system. In this case, you may want to recompile the software and optimize for the new architecture and its performance features. One highly valued compiler feature is the ability to optimize for the new architecture while maintaining compatibility with the old architecture.

Processor dispatch technology solves the problem of upgrading architectures while maintaining compatibility with legacy hardware deployed in the field. Intel has released a number of instruction set extensions such as MMX™ technology (Multimedia Extensions) and Streaming SIMD Extensions (SSE, SSE2, and SSE3). For example, the Intel Pentium 4 processor supports all of these instruction set extensions, but older processors, such as the Pentium III microprocessor, support only a subset of these instructions. Processor dispatch technology allows the use of these new instructions when the application is executing on the Pentium 4 processor and designates alternate code paths when the application is executing on a processor that does not support the instructions. There are three main techniques available that enable developers to dispatch code based upon the architecture of execution:

- Explicit coding of cpuid – programmer can add run-time calls to a function that identifies the processor of execution and call different versions of code based upon the results.

- Compiler manual processor dispatch – language support for designating different versions of code for each processor of interest. Figure 5.11 is a manual processor dispatch example that designates a code path when executing on Pentium III processors versus a code path when executing on other processors.

- Automatic processor dispatch – the compiler determines the profitability of using new instructions and automatically creates several versions of code for each processor of interest. The compiler inserts code to dispatch the call to a version of the function dependent on the processor that is executing the application at the time.

These processor dispatch techniques help developers take advantage of capabilities in newer processors while maintaining backward compatibility with older processors.

```
__declspec(cpu_specific(generic))
void array_sum(int *r, int *a, int *b,size_t l) {
  /* Some interesting function */
}
__declspec(cpu_specific(pentium_iii))
void array_sum(int *r,int const *a, int *b, size_t l) {
  /* Some interesting function optimized with SSE */
}
__declspec(cpu_dispatch(generic, pentium_iii))
void array_sum (int *r,int const *a, int *b, size_t l)) {
}
```

Figure 5.11: Manual dispatch example

The drawback to using these techniques is a code size increase in your application; there are multiple copies of the same routine, only one of which will be used on a particular platform. Developers are cautioned to be cognizant of the code size impact and consider only applying the technique on critical regions in the application.

5.3.3 Compile Time

Compile time is defined as the amount of time required for the compiler to complete compilation of your application. Compile time is affected by many factors such as size and complexity of the source code, optimization level used, and the host machine speed. The following sections discuss two techniques to improve compile time during application development.

5.3.4 PCH Files

Precompiled Header (PCH) files improve compile time in instances where a subset of common header files is included by several source files in your application project. PCH files are essentially a memory dump of the compiler after processing a set of header files that form the precompiled header file set. PCH files improve compile time by eliminating recompilation of the same header files by different source files. PCH files are made possible by the observation that many source files include the same header files at the beginning of the source file.

Figure 5.12 shows a method of arranging header files to take advantage of PCH files. First, the common set of header files are included from a single header file that is

```
/* global.h begin */               /* file2.cpp begin */
#include <iostream>                 #include "global.h"
#include <iomanip>                  #pragma hdrstop
#include <fstream>                  #include <vector>
#include <cstdlib>                  /* end of file2.cpp */
/* end of global.h */

/* file1.cpp begin */
#include "global.h"
/* end of file1.cpp */
```

Figure 5.12: Precompiled header file source example

included by the other files in the project. In the example below, `global.h` contains the include directives of the common set of header files and is included from the source files, `file1.cpp` and `file2.cpp`. The file, `file2.cpp`, also includes another header file, vector. Use the compiler option to enable creation of a PCH file during the first compilation and use the PCH file for the subsequent compilations. During compiles with automatic PCH files, the compiler parses the include files and attempts to match the sequence of header files with an already created PCH file. If an existing PCH file can be used, the PCH file header is loaded and compilation continues. Otherwise, the compiler creates a new PCH file for the list of included files. `#pragma hdrstop` is used to tell the compiler to attempt to match the set of include files in the source file above the pragma with existing PCH files. In the case of `file2.cpp`, the use of `#pragma hdrstop` allows the use of the same PCH file used by `file1.cpp`. Please note that the creation and use of too many unique PCH files can actually slow compilation down in some cases.

The specific compilation commands used for the code in Figure 5.12 when applying PCH files are:

```
icc -pch -c file1.cpp
icc -pch -c filc2.cpp
```

Try timing the compilations with and without the –pch option on the command line.

Table 5.9 shows compile time reductions for the code in Figure 5.12 and two other applications. POV-Ray [7] is a graphical application used to create complex images. EON

Table 5.9: Compile time reduction using PCH

Application at -O2	Compile time reduction (%) (Intel C++ Compiler, Version 9.0 Build 20050430Z)
Figure 5.12 Code	302
POV-Ray*	37
EON*	24

Table 5.10: Compile time reduction using make -j 2

Application at -O2	Compile time reduction (%) (Intel C++ Compiler, Version 9.0 Build 20050430Z)
POV-Ray	45
EON	48

[8], a graphical application, is a benchmark application in SPEC CINT2000. These tests were run on a 2.8 GHz Pentium® 4 microprocessor-based system with 512 MB RAM and Red Hat Linux*.

5.3.5 Parallel Build

A second technique for reducing compilation time is the use of a parallel build. An example of a tool that supports parallel builds is make (with the -j option). Table 5.10 shows the compile time reduction building POV-Ray and EON on a dual processor system with the make -j2 command. These tests were run on a dual 3.2 GHz Pentium® 4 microprocessor-based system with 512 MB RAM and Red Hat Linux.

5.3.6 Code Size

The compiler has a key role in the final code size of an embedded application. Developers can arm themselves with knowledge of several compiler optimization settings that can impact code size, and in cases where code size is important can reduce code size by careful use of these options.

Many compilers provide optimization switches that optimize for size. Two -O options optimize specifically for code size. The option -O1 optimizes for code speed and

Table 5.11: Code size related options

Optimization	Description
-O1	Optimize for speed, but disable optimizations that commonly result in increased code size.
-Os	Optimize for size, perhaps choosing instructions that result in smaller code size, but lower performance.
-unroll0	Disable loop unrolling.
-nolib_inline	Disable using optimized inlined version of common library functions.
-fno-exceptions	Disable generation of exception handling code.

code size, but may disable options that typically result in larger code size such as loop unrolling. The option -Os optimizes for code size at the expense of code speed and may choose code sequences during optimization that result in smaller code size and lower performance than other known sequences.

In addition to the -O options above that either enable or disable sets of optimizations, developers may choose to disable specific code size increasing optimizations. Some available options include:

- Options for turning off common code size increasing optimizations individually such as loop unrolling.

- Options to disable inlining of functions in your code or in libraries.

- Options to link in smaller versions of libraries such as abridged C++ libraries.

The last code size suggestion pertains to the advanced optimizations detailed in the first section of this chapter. Automatic vectorization and interprocedural optimization can increase code size. For example, automatic vectorization may turn one loop into two loops: one loop that can iterate multiple iterations of the loop at a time, and a second loop that handles cleanup iterations. The inlining performed by interprocedural optimization may increase code size. To help mitigate the code size increase observed with automatic vectorization or interprocedural optimization, use profile-guided optimizations that enable the compiler to use those optimizations only where they have the largest return. Table 5.11 summarizes a list of common code size optimizations. The option -fno-exception should be applied carefully in cases where C++ exception handling is not needed. Many embedded developers desire the use of C++ language features such as objects, but may not want the overhead of exception handling.

It is also good practice to "strip" your application of unneeded symbols. Some symbols are needed for performing relocations in the context of dynamic binding; however, unneeded symbols are eliminated by using (on Linux systems)

```
strip -strip-unneeded
```

5.3.7 Code Coverage

Many embedded developers employ a test suite of common inputs into their embedded application that provides some confidence that newly implemented code does not break existing application functionality. One common desire of a test suite is that it exercises as much of the application code as possible. Code coverage is functionality enabled by the compiler that provides this information. Specifically, code coverage provides the percentage of your application code that is executed given a specific set of inputs. The same technology used in profile-guided optimizations enables some code coverage tools.

Figure 5.13 is a screen shot of a code coverage tool reporting on a sample program. The tool is able to highlight functions that were not executed as a result of running the application input. The tool is also able to highlight areas inside an executed function that were not executed. The report can then be used by test suite developers to add tests that provide coverage. Consider the use of code coverage tools to increase the effectiveness of your test suite.

5.3.8 Optimization Effects on Debug

Debugging is a well-understood aspect of software development that typically involves the use of a software tool, a debugger that provides:

- Tracing of application execution
- Ability to stop execution at specified locations of the application
- Ability to examine and modify data values during execution

One of the trade-offs in optimizing your application is the general loss of debug information that makes debugging more difficult. Optimization effects on debug

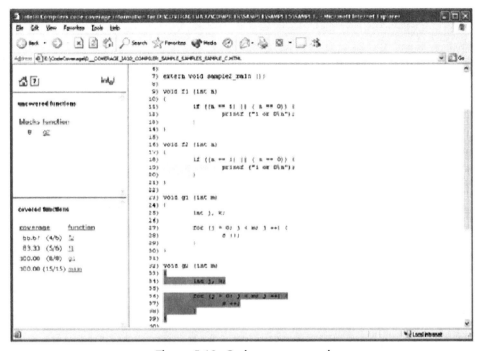

Figure 5.13: Code coverage tool

information are a large topic in itself. In brief, optimization affects debug information for the following reasons:

- Many optimized x86 applications use the base pointer (EBP) as a general purpose register which makes accurate stack backtracing more difficult.

- Code motion allows instructions from different lines of source code to be interspersed. Stepping through lines of source sequentially is not truly possible in this case.

- Inlining further complicates the code motion issue because a sequential stepping through of source code could result in the display of several different source locations leading to potential confusion.

Table 5.12 describes several debug-related options. These options are specific to the Intel C++ Compiler for Linux; however, similar options exist in other compilers. Please consult your compiler documentation to determine availability. The -g option is

Table 5.12: Debug related options

Optimization	Description
-g	Enable debug information generation.
-debug variable-locations	Better debug information for variables in optimized programs. Tracks variables whose particular value is represented by a register.
-debug inline-debug	Enables simulated call frames for inline functions.
-debug semantic-stepping	Instructs debugger to stop at the end of a machine instruction sequence that represents a source statement.

the standard debug option that enables the source information to be displayed during a debug session. Without this option, a debugger will only be able to display an assembly language representation of your application. The other options listed are used in conjunction with the -g option and provide the capabilities listed. Consider the use of supplemental debug options when debugging optimized code.

Chapter Summary

This chapter discussed several performance features and usability features available in many C++ compilers. Scalar optimization techniques are a necessary prerequisite before beginning parallel optimization. Understand and use the performance features and usability features available in your compiler to meet performance goals and increase developer efficiency. The net effect of using these features is cost savings for developing embedded applications.

Related Reading

Aart Bik has written a book on vectorization [9] and also authored a good introductory paper [10]. A dated book that serves as a good introduction into computer architecture is *High Performance Computing, Second Edition*, by Charles Severance and Kevin Dowd. Finally, the classic architecture book, *Computer Architecture, Fourth Edition: A Quantitative Approach*, by John Hennessy and David Patterson is a must-have for any serious software developer seeking to understand processor architecture.

References

[1]   SPEC 2000, www.spec.org

[2]   MySQL, http://www.mysql.com/

[3]   SetQuery Benchmark, http://www.cs.umb.edu/~poneil/poneil_setq/setq.tar.Z

[4]   ISO/IEC 9899:1990, Programming Languages C.

[5]   ISO/IEC 9899:1999, Programming Languages C.

[6]   ISO/IEC 14882, Standard for the C++ Language.

[7]   Persistence of Vision Raytracer, www.povray.org

[8]   EON, http://www.spec.org/cpu/CINT2000/252.eon/docs/252.eon.html

[9]   A. Bik, *The Software Vectorization Handbook*. Hillsboro, OR: Intel Press, 2004.

[10]   A. Bik, *Vectorization with the Intel® Compilers (Part I)*, http://www3.intel.com/cd/ids/developer/asmo-na/eng/65774.htm?page=1

## CHAPTER 6

Parallel Optimization Using Threads

Key Points

- Threads enable the utilization of multiple processor cores to complete a task; however, special handling is required to coordinate work both correctly and efficiently.

- The performance increase offered by parallelism is limited by the amount of the application that must execute serially. Several conditions limit the ability of code to execute in parallel. Some of these conditions can be removed, while some cannot.

- Focus analysis on the areas of highest execution as this will provide the highest return on your investment. Use the execution time profile to determine bounds on expected speedup from the parallelization effort.

- Thread-related bugs are difficult to diagnose because they are non-deterministic; the symptom may only show if threads execute in a specific order.

- Synchronization solves data race issues but can lead to thread stalls and deadlocks. Thread verification tools can help find thread-related bugs. When using thread verification tools, ensure the regions of code that are threaded are executed during the analysis run.

- Use thread performance tools to help gauge the amount of the application executing in parallel, the workload balance, and amount of synchronization occurring.

Accomplishing tasks in parallel comes in many guises. At the highest level, multiple computers can divide a task and work on it collectively. An example of this type of parallelism is the SETI@home [1] project where thousands of computers work collectively on analyzing radio signals for patterns indicative of intelligence. At the lowest level, modern microprocessors execute instructions specified in its machine language in parallel. For example, the Intel® Pentium® processor has a U execution pipe and a V execution pipe capable of executing different instructions at the same time. The motivation for enabling parallel execution at these different levels of computation is performance. Furthermore, the use of one type of parallel execution does not negate the benefit of employing other types of parallel execution. Threads are another means of enabling parallelism that complements the already mentioned types and is of smaller granularity than the process level concurrency mentioned in the SETI@home example, but certainly of larger granularity than the instruction level parallelism example of the Pentium® processor. Threads offer the ability to have multiple sections of one task in flight at the same time. In technical-speak: threads offer the ability to concurrently execute multiple contexts of the same process.

A real life analogy helps illustrate how threading relates to the other two types of parallelism. Consider the task of harvesting a crop from a farm and a worker as one tractor. The farm is split into multiple fields and a tractor only works on one field. In this example, harvesting the crop is the task (e.g., the work that needs completion). If there is only one tractor per field, there is no need to explicitly coordinate the work of the tractors. Each tractor has their pre-assigned space to harvest and can complete the task independently of the other tractors. Now, assume a second tractor is added to work one field. In this case the two tractors need to be concerned about what each other is doing. There are basic issues of feasibility – how do the tractors keep from running into each other? There are issues of optimizing performance – how is the work split between the tractors to maximize performance and harvest as quickly as possible? This real life example is similar to threading. The threads are akin to multiple tractors tending to one field and similar issues exist. Threads must coordinate their work properly and efficiently or risk crashes and low performance.

In the previous example as long as there was only one tractor on each field they were safe. This is similar to the operating system (OS) concept of a process. Two processes do not share the memory that the OS assigns to them[1] so the chances of them interfering with each other are minimized. In terms of scheduling the processes, the OS determines where and when each process will execute. With regard to threads, the software that

---

[1] Processes physically share the memory on the system, but the OS makes it appear to the process to be disjoint memory regions. Modern OSes will throw an exception if a process attempts to access memory outside of its assigned memory.

is either implemented by the programmer or provided by a library that abstracts some of this low level work must coordinate the threads. Nevertheless, designing for threads requires a disciplined process for best effect.

In order to take advantage of threading, the software development process detailed in this chapter is recommended. The *Threading Development Cycle* (TDC) focuses on the specific needs and potential challenges introduced by threads. It is comprised of the following steps:

1. Analyze the application

2. Design and implement the threading

3. Debug the code

4. Performance tune the code

This chapter is specific to cases where the OS provides software threads and shared memory although the steps detailed in the TDC can be applied in the more general case. Before taking a deeper look at the TDC, some basic concepts of parallel programming are explained.

6.1 Parallelism Primer

The typical goal of threading is to improve performance by either reducing latency or improving throughput. Reducing latency is also referred to as reducing turnaround time and means shortening the time period from start to completion of a unit of work. Improving throughput is defined as increasing the number of work items processed per unit of time.

6.1.1 Thread

A thread is an OS entity that contains an *instruction pointer*, stack, and a set of register values. To help in understanding, it is good to compare a thread to a process. An OS *process* contains the same items as a thread such as an instruction pointer and a stack, but in addition has associated with it a memory region or heap. Logically, a thread fits inside a process in that multiple threads have different instruction pointers and stacks, but share a heap that is associated with a process by the OS.

Threads are a feature of the OS and require the OS to share memory which enables sharing of the heap. This is an important fact because not all embedded OSes support shared memory and as a result, not all embedded OSes support threads. A clarification is that the type of threads discussed in this chapter is user level software threads. Hardware threads are a feature of many microprocessors and are quite distinct in the meaning and capability. For example, the term simultaneous multi-threading is a microprocessor feature that enables one processor core to appear and function as multiple processor cores. For the purposes of this chapter, hardware threads are relevant only in that they may provide the processor cores that the software threads execute upon.

One other clarification is that this chapter discusses user level threads only. We do not include discussion of kernel level threads or how different OSes map user level threads to kernel threads.

6.1.2 Decomposition

Effectively threading an application requires a plan for assigning the work to multiple threads. Two categories of dividing work are *functional decomposition* and *data decomposition* and are summarized as:

1. Functional decomposition – division based upon the type of work.

2. Data decomposition – division based upon the data needing processing.

Functional decomposition is the breaking down of a task into independent steps in your application that can execute concurrently. For example, consider an intrusion detection system that performs the following checks on a packet stream:

- Check for scanning attacks

- Check for denial of service attacks

- Check for penetration attacks

As long as each step above was an independent task, it would be possible to apply functional decomposition and execute each step concurrently. Figure 6.1 shows sample OpenMP code that uses the section directive to express the parallelism.

When this code is executed, the OpenMP run-time system executes the function in each OpenMP section on a different thread and in parallel (as long as the number of processor cores exceeds the number of threads executing).

In practice, attempting to execute multiple threads on the same data at the same time may result in less than ideal performance due to the cost of synchronizing access to the data.

```
#pragma omp parallel sections
{
#pragma omp section
  Check_for_scanning_attacks() ;

#pragma omp section
  Check_for_denial_of_service_attacks() ;

#pragma omp section
  Check_for_penetration_attacks() ;
}
```

Figure 6.1: Functional decomposition example

```
#pragma omp parallel for
{
for (i=0;i<1000000;i++) {
  process_image(i);
}
```

Figure 6.2: Data decomposition example

Pipelining is a category of functional decomposition that reduces the synchronization cost while maintaining many of the benefits of concurrent execution. Chapter 8 contains a case study that employs pipelining to enable parallelism.

Data decomposition is the breaking down of a task into smaller pieces of work based upon the data that requires processing. For example, consider an image processing application where multiple images need to be converted from one format to another format. Each conversion takes on the order of seconds to complete and the processing of each image is independent of the processing of the other images. This application lends itself quite naturally to data decomposition. Figure 6.2 shows sample OpenMP code using the parallel for directive.

When this code is executed, the OpenMP run-time system will divide the processing of the images between the allocated threads for parallel execution (assuming the number of processor cores is greater than 1). If the processing of each individual image consumed a great deal of time, it may make sense to multi-thread the processing of the individual image and execute the processing of the subimages by different threads. Chapter 7 presents a case study that employs data decomposition in order to multi-thread image rendering.

In general, it is easier to scale applications by employing data decomposition than it is using functional decomposition. In practice you may find a combination of different decompositions works well for your particular application.

6.1.3 Scalability

Scalability is the degree to which an application benefits from additional processor cores. As the number of cores the application uses is increased, it would be nice if performance of the application increased as well. There are natural limits to how much of an application can be executed in parallel and this limits the obtained performance benefit of parallelism. Amdahl's Law is used to compute the limits of obtained performance and is expressed [2]:

$$\text{Speedup} = \cfrac{1}{(1 - \text{Fraction}_e) + \cfrac{\text{Fraction}_e}{\text{Speedup}_e}}$$

where Fraction_e = the amount of the application that executes in parallel; and Speedup_e = how many times faster the parallel portion executes compared to the original.

For example, consider an application that executes in parallel 50% of the time. Also, assume the application executes on a system with four processor cores and was threaded in such a way that the performance scales with the number of processor cores. In this example, $\text{Fraction}_e = 0.5$ and $\text{Speedup}_e = 4$, and therefore Speedup = 1.6.

Efficiency is a measure of how effectively the total number of processor cores is being employed in running the application in parallel; the goal is a 100% measure of efficiency. In the previous example, three processor cores are idle for 4/5 of the execution time, which means 12/20 of the four processor cores' time is idle and thus 8/20 of the time the processor cores are active with 40% efficiency.

Consider another example involving the aforementioned image processing problem with the following constraints:

- Ten seconds of initialization that must run serially

- One second to process one image (processing of different images can be accomplished in parallel)

- Ten seconds of post-processing that must run serially

Table 6.1 shows the calculations of scalability and efficiency for a number of different processor cores. One observable trend is that as the number of processor cores increases, the corresponding decrease in execution time is not as significant. In other words, the speedup does not scale linearly with the number of processor cores. For example, the scalability with 32 processor cores is 10.89 and with 300 processors is 15.24, not 108.9 (10 × 10.89). This trend occurs because the serial portion of the application is beginning to dominate the overall execution time. With 16 processor cores, the execution time of the image processing step is 300/16 = 18.75 s. With 300 processor cores, the execution time of the image processing step is 300/300 = 1 s. Furthermore, 299 processor cores are active for only 1 s out of the 21 s of total execution time. Thus, efficiency is 5.1%. The conclusion is: maximize the benefits of parallelism by parallelizing as much of the application as possible.

One other point worth mentioning is that scalability should always be compared against the best achievable time on one processor core. For example, if the use of a new compiler and optimization settings resulted in a decrease in execution time when run on one processor core, the scalability numbers and efficiency percentages should be recalculated.

6.1.4 Parallel Execution Limiters

All problems do not lend themselves to parallelization and understanding some common limiters is essential. Also, some problems may be executed in parallel only after special

Table 6.1: Image processing scalability and efficiency

Processor cores	Initialization	Image	Post-processing	Total time	Scalability	Efficiency (%)
1	10	300	10	320	1.00	100
2	10	150	10	170	1.88	94.1
4	10	75	10	95	3.37	84.2
8	10	37.5	10	57.5	5.57	69.6
16	10	18.75	10	38.75	8.26	51.6
32	10	9.375	10	29.375	10.89	34.0
300	10	1	10	21	15.24	5.1

```
for (i=0;i< num_steps; i++){
    x = (i+0.5) * step;
    sum = sum + 4.0/(1.0 + x*x);
}
```

Figure 6.3: Data dependency example

handling of these potential limiters. The key limiters to efficient parallel execution are summarized as:

1. Shared variables – variables that are accessible by multiple threads. Typically, these are global variables or local variables in scope of two or more threads.

2. Data dependency – variable A is said to be data dependent upon variable B if the value of A is calculated based in part or whole upon the value of variable B.

3. State in routines – routines with values saved and used across invocations.

4. Memory bandwidth capacity – applications executing in parallel are demanding more data than the path to memory can serve at one time.

The loop in Figure 6.3 has a data dependency between iterations on the sum variable and would typically limit executing the code in parallel.

Suppose the loop in Figure 6.3 was threaded by dividing the iteration space between two threads. Thread one executes the loop from *i* equal to 0 to (num_steps/2)−1 and thread two executes from *i* equal to num_steps/2 to num_steps−1. In order to ensure correct parallel execution, the variables referenced inside of the loop body require special handling. As an illustration, consider the threading impact on the variable, *sum*. For simplicity, assume that (4.0/(1.0+x * x))=1.0 and that *sum* is initially 0. Figure 6.4 shows a potential timeline of execution for the two threads as they execute the loop in parallel. In this figure, the C statement, sum=sum+4.0/(1.0+x * x), is represented by a read operation, a register operation, and a subsequent write operation. The read operation takes the value from memory and places it into a register. The write operation takes the value that was computed in the register and places it back into memory. Please keep in mind that the register used by thread one and thread two are distinct. The execution of the C statement is not guaranteed to be atomic, which means that thread one may be in the middle of reading and writing when thread two starts its own read and write.

Time	Thread 1	Thread 2
0	Read sum (sum = 0) into register	
1	Compute using register	Read sum (sum = 0) into register
2	Write sum (sum = 1) from register to memory	Compute using register
3		Write sum (sum = 1) from register to memory

Figure 6.4: Data race timeline

```
for (i=0;i< num_steps; i++){
    sum = sum + extra;
    extra = i;
}
```

Figure 6.5: Data dependency example 2

From Figure 6.4, you see that the sum variable is incremented two times; however, at time 3 after the two increments, sum is still equal to one which is obviously incorrect. This simple example illustrates one of the fundamental challenges when threading applications – the determination of which variables need to be shared between threads and how to effectively coordinate access to them. To add to the difficulty of this challenge, there is no guarantee that the problem detailed above will actually exhibit itself every time you execute the application. The execution order of threads is non-deterministic. This means that the problem may not occur during the development and subsequent debug of the code and only rears itself in the field once a customer is using the application. In other words, some of the time thread one will read and write the sum variable uninterrupted as will thread two. This issue is termed a race condition and techniques to handle this type of issue are detailed later in this chapter.

A more complicated example is shown in Figure 6.5.

The complication is due to the dependency between the variable sum and the variable extra that spans across iterations of the loop. A method of testing if a particular loop can be parallelized is to run the loop backward in terms of its loop indexes. In this simple example, *sum* will have a different value if executed with `i=num_steps-1` to 0 versus `i=0` to `num_steps-1`. While this example is conceptually simple, the issue manifests itself in many types of code, particularly those involving matrix operations. In addition, some forms of hand optimizations commonly performed such as loop unrolling and software pipelining can introduce these dependencies and make threading problematic.

An example of state in routines is a routine that maintains data between invocations of the routine using statically defined variables. Examples of routines with state include typical memory allocation routines, random number generators, and I/O routines. Three solutions to calling functions that are not thread safe in your parallel code are:

1. Restrict access to these functions to allow only one thread to call the function at a time.

2. Use thread safe versions of the routines.

3. Develop your routines to be reentrant.

Many library providers ensure thread safety as a feature of their library. Consult your library's documentation to ensure that they are thread safe before using them in your multi-threaded code.

Routines that are made reentrant are capable of being executed in parallel. These routines would not have dependencies between other invocations of the same routines. A typical sign that a routine is not reentrant is if the routine defines and uses *static variables*.

Global variables pose a challenge because the values are shared between the threads. First, there may be correctness issues once different threads access these values. The correctness issues are handled using one of two methods:

1. Minimize the use of global variables.

2. Use synchronization to limit access to each global variable.

Unnecessary use of global variables is characteristic of older software developed before modern software engineering practices. Reducing the use of global variables in these cases can reduce the need for synchronization.

Memory bandwidth can also limit performance of multi-threaded applications because the bus is a shared resource between the processor cores. If the threads and other applications on the system are memory demanding with many reads from and writes to large amounts of data, it is possible for the bus to become the bottleneck. This is a difficult issue to diagnose; symptoms that this issue may be occurring include:

- Central processing units (CPU) utilization is close to 100%

- Poor performance scaling is observed moving to a four or eight processor core system

- Low number of context switches per second

It is also possible to compare the performance of your multi-threaded application when executing one thread with the application executing many threads and obtain an event-based sampling (EBS) profile on clockticks for both. Hotspots in the profile with many threads that do not occur in the profile with one thread may be indicative of high bandwidth demanding portions of your code.

Intel Architecture processors also have performance monitoring counters (PMCs) with events to measure usage of the bus. Information on employing the PMCs to determine if the bus is saturated in this manner can be found in an article by Levinthal [3].

6.1.5 Threading Technology Requirements

Chapter 5 detailed several technologies in use today to thread applications and these technologies were classified into two broad categories, library-based and compiler-based threads. All of these threading technologies place requirements on the software development environment, the host and target system, and the end applications.

Library-based thread application programming interfaces (APIs) like POSIX threads or Win32 threads require the linking in of these thread libraries into your code. During compilation, the source files will reference the interfaces through the thread library header files. Your source code can make calls to routines in the library and therefore during the link phase, the appropriate thread library must be referenced. The steps are:

- Source code will contain include directives referencing the thread API header files

- Source code will contain the appropriate calls to the thread API routines that are used

- The thread API routines will be linked in during the link phase either dynamically or statically

These steps are no different than using any other type of library; however, the impact is worth detailing. The library routines that are linked into the application will add to the code size of your application in the case of static linking. In the case of dynamic linking the shared object will need to be available and supported on your target system.

Domain-specific thread libraries often rely upon lower level thread libraries. As a result, use of domain-specific libraries can require access to the underlying lower level thread library as well as the domain-specific library. For example, the multi-threading routines available in Intel® Integrated Performance Primitives (IPP) rely upon POSIX threads on Linux systems. Therefore, an application that makes calls to multi-threaded Intel® IPP routines may have dependencies on Pthreads as well as Intel® IPP. The situation with compiler-based threading is similar to domain-specific libraries. In addition to requiring headers and libraries provided by the compiler, there may be dependencies on underlying thread libraries when features such as autoparallelization and OpenMP are used.

6.2 Threading Development Cycle

Embedded application development is aided by two components, the right development tools and the right development process. This section details a threading development process that can aid developers employing threads in their application. The TDC is iterative and concludes when the application requirements have been met. Figure 6.6 depicts the TDC which is summarized by these four steps:

1. Analysis – determining which portion of the application to thread.

2. Design – deciding how to thread and implementing.

3. Debug – finding and fixing thread-related stability issues in the code.

4. Tune – optimizing for thread performance issues.

The next four sections provide details on each of the above steps.

6.2.1 Analysis

The first step of the TDC is analysis, which seeks to determine which portions of the code to thread. Threading should be applied to the portions of the application that execute the most frequently and will thus lead to the largest performance improvement. Alternatively,

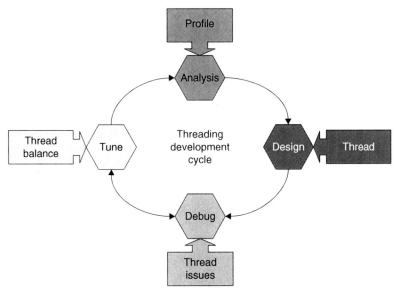

Figure 6.6: Threading development cycle

there is no point in threading portions of the code that are never or rarely executed in practice. So the question of where to thread is important. The analysis phase can be broken down into the following steps:

1. Develop benchmark that is representative of the application

2. Tune for serial performance

3. Obtain an execution time profile

4. Identify top regions of code and top loops

5. Obtain a call graph profile

6. Create a flow chart of the critical algorithms

6.2.1.1 Benchmark

Before the question of where to thread can be answered, a reproducible workload, or benchmark, of the application should be created. In order to conduct the analysis and successive steps of the TDC, the benchmark should have a number of critical qualities summarized as follows:

- Representative of common usage patterns

- Reproducible

- Simple to execute

- Reduced time to execute

First, the benchmark should represent typical usage patterns of the application by the user. For example, the benchmark should not be a regression test suite designed to test every possible code path through the application. Second, steps to exercise the benchmark should be reproducible so that the effects of the threading implementation can be analyzed for both stability and performance. Third, the benchmark should be simple to execute to minimize the chance for human error. Finally, the benchmark should complete in a short duration which enables quick turnaround on performance experiments.

6.2.1.2 Tune for Serial Performance

Serial tuning is less labor intensive and can lead to dramatic performance improvements that can enhance the benefits of parallel execution if done properly. There is a caveat; some serial optimization can limit the ability to parallelize. The best way to consider if over-optimization may inhibit parallelism is to determine if your optimizations are adding the elements that inhibit parallelism defined in Section 6.1.4.

6.2.1.3 Collect an Execution Time Profile

An execution time profile is just as critical for threading as it is for standard scalar optimization. The execution time profile is a report on the areas of highest execution of an application. Threading these frequently executed regions of code typically provides the largest performance improvement. The benefits of obtaining the execution time profile are no different than what was detailed in Section 4.3.7.1. In addition, with this information it is possible to calculate theoretical speedup as a result of parallel execution. For example, suppose an application has the execution time profile shown in Table 6.2. Fifty percent of the execution time occurred in the function, `heavy_compute2()`. If you planned to execute the application on a machine with four processor cores and were able to achieve linear scaling, the expected speedup for the entire application is 1.6. This estimate is

Table 6.2: Execution time profile example

Function	Time in function (in seconds)	Estimated time in parallel version (in seconds)	Estimated time in parallel version 2 (in seconds)
main()	0.5	0.5	0.5
heavy_compute1()	3.5	3.5	0.875
heavy_compute2()	5.0	1.25	1.25
heavy_compute3()	1.0	1.0	0.25
Total	10.0	6.25	2.875

helpful in understanding the return on investment of the effort. If the project requirements could be met by threading `heavy_compute2()`, you may choose not to thread other regions. If threading of `heavy_compute2()` did not allow you to reach your project requirements, you could consider threading of `heavy_compute1()` and `heavy_compute3()`. In addition, you can also communicate what speedup is reasonably possible. If management demanded a 300% speedup, but did not allocate the resources necessary to thread `heavy_compute1()` and `heavy_compute3()`, you can clearly communicate the infeasibility of the plan.

6.2.1.4 Collect a Call Graph Profile

A call graph profile helps determine where in the application to introduce threads. A call graph shows relationships between a function (caller) and the function it calls (callee). A typical application can contain thousands of functions with many levels or nests of calls and as a result a call graph of the application can be very complex. A useful call graph will contain additional information such as the number of times a function is called, the amount of time spent in the function, and calculations of the most time-consuming paths through the call graph.

A call graph benefits analysis of your application in two areas:

1. Easily identify portions of the application that may be amenable to threading via functional decomposition.

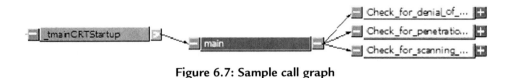

Figure 6.7: Sample call graph

2. Gain a better understanding of the affects that a threading decision in one area has on the larger application.

Figure 6.7 is a sample call graph created by profiling code similar to the code in Figure 6.1. One possibility observed from the call graph is that the three functions called from main may be candidates for parallel execution.

On the second point, a call graph provides an idea of which functions could be affected by the area chosen to be threaded. For example, if you decided to thread a routine that called several other functions which in turn called many others, it may be good to do a cursory analysis of the variables used in the routines to determine if access to shared variables needs to be synchronized. Many synchronization issues could be prevented upfront which saves time during the debug phase.

6.2.1.5 Flow Chart Hotspots

A flow chart of hotspots is useful in cases where the call graph does not reveal the functional flow of algorithm steps with enough granularity. Furthermore, in flow charting the algorithm, it should become clear if functional or data decomposition should be employed. Producing a flow chart is a time-intensive activity that should be employed on only the hotspots. Figure 6.8 shows a sample flow chart of 179.art, one of the SPEC CPU2000 [4] benchmarks that implements a pattern recognition application employing an Adaptive Resonance II [5] neural network, an algorithm which is very amenable to data decomposition. In fact, a threaded version of the pattern recognition application is used in the SPEC OMP [6] benchmarks and is known as 330.art_m and 331.art_l. The OpenMP version of the benchmark [7] is parallelized based upon data decomposition of the outer loop shown in Figure 6.8. The benefit of the flow chart is that it visually supplements knowledge of frequently executed loops and a call graph into a diagram that aids in your multi-thread design.

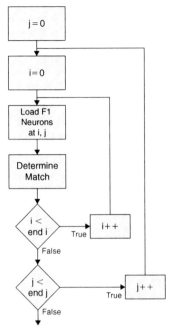

Figure 6.8: Flow chart of scanner application

6.2.1.6 Classify Most Frequently Executed Loops

The next step in analysis is to classify the most frequently executed loops in the application. Loop level parallelism via data decomposition is typically an easier undertaking than other types of parallelization and should be considered first. The term, top loop, is relative to the application and your performance needs, but typically 10% or more of the execution time is worthy of consideration. In analyzing loops for threading focus on both innermost loops with execution time concentrated in a small area and loops with large bodies and potentially many nested loops within. The initial focus on both small and large loops is a difference compared to typical performance analysis which focuses first on tight innermost loops. A greater variety of loops is considered because the cost of spawning threads should be amortized at the correct level or it is possible to realize less than ideal benefits from threading. For example, consider the code in Figure 6.9. If the function, `perform_task()`, was threaded using Pthreads, the overhead of the thread creation and thread deletion would be contained inside the function call and executed every loop iteration. The thread overhead would negatively

```
for (i=0;i< num_steps; i++){
  perform_task(i);
}
```

Figure 6.9: Sample code before thread pool

```
for (i=0;i< num_steps; i++){
  QueueUserWorkItem(perform_task, (void *)&I, 0);
}
```

Figure 6.10: Sample code using thread pool

```
for (i=0;i< num_steps; i++){
  for (j=0;j < (num_steps * 2); j++) {
    do_work();
  }
}
```

Figure 6.11: Classify most frequently executed loop

impact performance. A better approach is detailed in Figure 6.10 using a thread pool which incurs the overhead of thread creation and deletion only once over the iterations of the loop. The thread pool used in this example follows Microsoft syntax; however, the thread pool pattern is not unique to it. Previous to the call to QueueUserWorkItem, a function call to setup the pool of threads will have been made allowing the loop to take advantage of a pool of threads without the setup and teardown overhead of thread creation and deletion.

It is not trivial to find the most frequently executed loops in an application. First of all, the definition of most frequently executed loop is ambiguous. It could mean the loop with the largest number of iterations or it could mean the loop containing the largest amount of execution time. For the purposes of performance analysis, most frequently executed loop implies both definitions. Consider the code in Figure 6.11 which contains a doubly nested loop.

The execution time of the outer loop is larger than the inner loop. The inner loop iterates more than the outer loop. For the purposes of threading code, both loops should be considered. In addition, not all loops are as well structured as a simple for loop. Loops could occur over large areas of code, use more complicated control structures such as while loops or even goto statements, and contain multiple entries into and exits out of the loop. The current generation of performance analysis tools do not contain a direct method of determining the top loops in the program. Instead you should analyze the code supplemented by the analysis information collected in the previous steps to determine the top loops and classify them appropriately. The following are questions to consider when determining the top loops:

- What does the flow chart tell you?

- What are the top functions in the application?

- What are the innermost loops in the function?

- In the call graph, what is the ratio between a callee function and a called function contained inside a loop?

- Do the innermost loops contain relatively the largest number of events in an execution profile?

Once the top loops are determined, the loops should be classified on several categories:

- What is the average execution count of the loop?

- Is the loop iteration count bounded upon entry into the loop?

- How well structured is the loop? Multiple entries and exits?

The average execution count of a loop can be estimated based on a number of ways. Two simple approaches are observing the loop using a debugger and adding output commands to display the loop iteration counts. The average execution count is important for a number of reasons. First, if the iteration count is small it may not make sense to parallelize the region. Second, a large iteration count will guide location of the threads that are created. Creating threads inside such a loop would create a large amount of overhead.

Some threading techniques require that a loop's iteration count is known before entry into the loop. The iteration count can be dynamically determined at run-time, but not dynamically altered once the loop has begun iterating. Some threading techniques require well-structured loops. If the loop has multiple entries and exits, it may prove difficult to thread the loop without significant modifications.

Table 6.3: Comparison of thread technologies

Category	Explicit threading (Pthreads)	OpenMP	Autoparallelism and speculative precomputation	Domain-specific library	Intel Threading Building Blocks
Flexibility	High	Moderate	Moderate	Specific to problem area	High
Ease-of-use	Low	Moderate	High	High	Moderate
Decomposition	Both	Data, some functional	Data	Data	Both
Portability	Moderate	Depends on compiler support	Depends on compiler support	Low	Moderate
Code impact	High	Low	Low	Moderate	Moderate

6.2.2 Design and Implementation

Designing for threads depends on a number of constraints including thread technology, application and target characteristics, customer requirements, and programmer preferences, to name a few. Chapter 4 summarized the available tool technology. It is out of the scope of this book to teach how to design for threads using every technology available. Instead, this section provides a comparison of different thread technology that will help guide your decision on which technology to implement on. The end of this chapter includes and summarizes a number of references to help learn different thread technology in detail.

Part of the design decision is the choice of technology on which to implement. Table 6.3 provides a high level comparison of various thread technologies.

Low level threads embodied by Pthreads tend to be very flexible in handling different parallelism requirements because the programmer controls the threads explicitly. The programmer is responsible for marshalling the code that needs to be threaded into a routine, explicitly calling the Pthreads routine to spawn the threads, handling communication and synchronization between threads, and destroying the threads. Flexibility comes at a price as the level of complexity is higher than other thread technologies. Explicit threading can be employed to perform both functional and data decomposition although using it for data decomposition can be challenging. Portability of Pthreads is widespread with many modern OSes supporting it; however, there are many embedded OSes that do not. The code impact is somewhat high because the modifications required to multi-thread the code results in code that looks very different than the original serial code. Choose explicit threads if maximum flexibility is desired and you are willing to invest extra effort compared to other thread technologies.

OpenMP is flexible in handling data decomposition very well, but is somewhat limited in handling functional decomposition. OpenMP sections can be used to multi-thread some programs via functional decomposition; however, the semantics require the number of functions and name of each function to be statically known. OpenMP is of medium difficulty as the programmer needs to understand where the threading occurs as well as denoting which variables are shared and which are private. The availability of compiler-based threading is dependent on compiler support and the underlying thread support. One advantage of OpenMP is that the code impact can be low. The parallel version of the code can look very similar to the original serial code if implemented correctly. An example of OpenMP's low impact on source code is discussed in the next section. Choose OpenMP if your algorithm is amenable to data decomposition.

Autoparallelism and speculative precomputation are similar to OpenMP in many of the categories listed in Table 6.3 except two. Ease of use is high because the user is only required to specify a compiler option on the command line; the compiler does the rest. Flexibility is lower than OpenMP in that the compiler is typically more conservative in threading regions of code due to the lack of programmer knowledge about program characteristics. Choose autoparallelism and speculative precomputation if you desire easy access to parallel execution and are willing to accept the limitations.

Domain-specific thread libraries, an example being Intel® IPP, are flexible within the particular problem domain. Intel® IPP contains threaded routines to perform many kinds of transforms specific to media processing, however, if Intel® IPP does not contain a function that meets your exact needs and your needs are not decomposed to lower level Intel® IPP functions, then you are out of luck. Choose domain-specific thread libraries if your needs can be expressed using the available library functions.

Intel® Threading Building Blocks (TBBs) is a fairly recent approach for parallelism. It is very flexible in handling both data and functional decomposition well. Ease of use is moderate requiring understanding of C++ templates. Portability is limited being restricted to Intel platforms and specific OSes. Code impact is moderate due to the need

```
#include <stdio.h>
#include <omp.h>
static int num_steps;
double step;
#define NUM_THREADS 2
int main ()
{
  int i;
  double sum = 0.0;
  omp_set_num_threads(NUM_THREADS);
  num_steps = omp_get_num_threads() * 10000;
  #pragma omp parallel for
  for (i=0;i< num_steps; i++){
    sum = sum + 4.0/(1.0+i);
  }
  printf("%lf\n", sum);
}
```

Figure 6.12: Code sample with unprotected OpenMP usage

to modify the serial code to call TBB template functions and use special TBB provided objects. Choose Intel® TBB if you are programming in C++ and need both data decomposition and functional decomposition.

6.2.2.1 Code Modifications

Once the design is ready, the code modifications are made. Traditional best-known software development practices should be employed such as using source control tools, code reviews, etc. One tip when implementing is to code a non-threaded version alongside the threaded version. The form of the non-threaded version will be different depending on the particular thread technology applied. For example, one strength of OpenMP is the low impact to the original serial source code. If coded properly, your source code that has been multi-threaded using OpenMP pragmas can be turned back into serial code by simply omitting the OpenMP option from the compiler command line. OpenMP has this ability because the default C behavior when compiling an unrecognized pragma is to emit a warning and continue compilation. Let's observe how to write OpenMP code so that it can execute correctly without the OpenMP option. First, Figure 6.12 contains code that has been multi-threaded using OpenMP that has two issues that prevent running serially:

```
#include <stdio.h>
#include <omp.h>
static int num_steps;
double step;
#define NUM_THREADS 2
int main ()
{
  int i;
  double sum = 0.0;
#ifdef _OPENMP
  omp_set_num_threads(NUM_THREADS);
  num_steps = omp_get_num_threads() * 10000;
#else
  num_steps = 10000;
#endif
  #pragma omp parallel for
  for (i=0;i< num_steps; i++){
    sum = sum + 4.0/(1.0+i);
  }
  printf("%lf\n", sum);
}
```

Figure 6.13: Sample OpenMP code with protected OpenMP usage

1. Unguarded call of OpenMP library routines.

2. Correct execution depends upon OpenMP library return value.

Figure 6.13 lists a modified version of the code in Figure 6.12 that is easily converted back and forth between a serial and OpenMP version.

The call to `omp_get_num_threads` is protected via an `#ifdef` directive and is only called when OpenMP is specified on the command line. The OpenMP specification states that the symbol _OPENMP should be defined when OpenMP is specified on the command line. The value, `num_steps`, is dependent on the return value of `omp_get_num_threads()`. A solution is to modify the code so no critical values are dependent on the OpenMP API function calls. One technique is to use an #else directive as an alternative compilation path when OpenMP is not specified on the command line.

Now, let's consider the compilation and link sequence and code size and dependency impacts. A typical compilation sequence for the code in Figure 6.13 is:

```
icc -c code.c -openmp
icc -o code code.o -openmp
```

Executing the size command on the executable reveals the following:

```
size code
```

Text	Data	Bss	Dec	Hex	filename
4492	560	32	5084	13dc	code

Without the -openmp option on the command line, the size command returns the following:

Text	Data	Dss	Dec	Hex	filename
3195	296	24	3515	dbb	code

As you can see, the use of the OpenMP option results in a larger executable. Please note that this example is a small program that uses a small set of OpenMP features. A typical application is much larger and the additional code size attributed to OpenMP will be amortized over the overall size of the application. Nevertheless, this size increase should be understood in the context of your embedded project.

Performing an ldd shows the libraries that the implementation is dependent upon:

```
ldd openmp
libm.so.6 => /lib/tls/libm.so.6 (0×40028000)
libguide.so => /opt/spdtools/compiler/ia32/cc-9.1.046/lib/
  libguide.so (0×4004a000)
libgcc_s.so.1 => /lib/libgcc_s.so.1 (0×40088000)
libpthread.so.0 => /lib/tls/libpthread.so.0 (0×40090000)
libc.so.6 => /lib/tls/libc.so.6 (0×42000000)
libdl.so.2 => /lib/libdl.so.2 (0×4009d000)
/lib/ld-linux.so.2 => /lib/ld-linux.so.2 (0×40000000)
```

As you can see, the use of OpenMP is dependent on a library provided by the compiler, libguide.so and the underlying Pthreads on the system, libpthread.so.0. In a cross-compilation environment, you should make sure all of the shared libraries are available on your target.

One other option on creating single source serial/parallel code is to use the OpenMP stubs library. To do this, compile the code in Figure 6.13 as you did before, but on the link line use `-openmp_stubs`. The code modification to handle the call to `omp_get_num_threads()` would become unnecessary because the stub libraries provide an implementation that returns 1. In general the stub routines available in the library behave with serial semantics.

The concept of using preprocessing to enable single source serial/parallel code works the same using other thread technologies. Some Pthreads implementations provide a stub library interface that can be used for the same purpose.

6.2.3 Debug

Once the code is implemented, typically it fails due to bugs in the program. It is rare to find complex code that is written correctly the first time and requires no debugging. The same is true for multi-threaded code. In fact, debugging multi-threaded code can be more challenging due to the insidious nature of the problems that can occur once code is executing in parallel. The developer is faced with a standard debug cycle that includes potential new bugs associated with the added threads. The debug phase seeks to answer the following questions:

- What non-thread-related bugs exist in the code and how will they be addressed?

- What thread-related bugs exist in the code? What kind of bugs? How will they be addressed?

The non-thread-related bugs can be handled using common development and debug techniques outside of the scope of this book.

6.2.3.1 Basic Multi-threaded Debugging

The debuggers referenced in Chapter 4 offer differing capabilities for assisting with the debug of multi-threaded code. At a minimum, the debuggers enable the switching between threads and examination of thread state during the debug session. Typical techniques that aid during multi-threaded debug include:

- Prevent multi-threaded bugs by using standard software development practices

- Implement logging or trace buffers as part of the debug path in the code

- Use thread-specific verification tools

Information on standard software development practices are available from many sources and will not be detailed here. Standard software development practices include the use of source control tools, integrated development environments, code reviews, and many other practices that reduce the frequency of software bugs. Before going into detail into the other two thread-specific debug techniques, basic thread-related bugs are discussed.

6.2.3.2 Thread-related Bugs

Thread-related bugs are difficult to detect and require extra time and care to ensure a correctly running program. A few of the more common threading bugs include:

- Data race

- Thread stall

- Deadlock

Data Race

A data race occurs when two or more threads are trying to access the same resource at the same time. If the threads are not synchronizing access, it is not always possible to know which thread will access the resource first. This leads to inconsistent results in the running program. For example, in a read/write data race, one thread is attempting to write to a variable at the same time another thread is trying to read the variable. The thread that is reading the variable will get a different result depending on whether or not the write has already occurred. The challenge with a data race is that it is non-deterministic. A program could run correctly one hundred times in a row, but when moved onto the customer's system, which has slightly different system properties, the threads do not line up as they did on the test system and the program fails. The technique to correct a data race is to add synchronization. One way to synchronize access to a common resource is through a critical section. Placing a critical section around a block of code alerts the threads that only one may enter that block of code at a time. This ensures that threads will access that resource in an organized fashion. Synchronization is a necessary and useful technique, but care should be taken to limit unnecessary synchronization as it will slow down performance of the application. Since only one thread is allowed to access a critical section at a time, any other threads needing to access that section are forced to wait. This means precious resources are sitting idle, negatively impacting performance.

```
#pragma omp parallel
{
  double psum = 0.0;
  #pragma omp for
  for (i=0;i< num_steps; i++){
    x = (i+0.5)*step;
    psum = psum + 4.0/(1.0+x*x);
  }
  #pragma omp critical
  sum += psum;
}
```
Figure 6.14: Reduce synchronization example

```
#pragma omp parallel for reduction(sum: +)
  for (i=0;i< num_steps; i++){
    x = (i+0.5)*step;
    sum = sum + 4.0/(1.0+x*x);
  }
```
Figure 6.15: OpenMP reduction example

Another question to ask before adding synchronization is if the variable in question really needs to be shared. If you applied an OpenMP parallel for directive in the example in Figure 6.3 without dealing with the data dependency on the sum variable, you would create an instance of a data race. Fixing this issue is accomplished by synchronizing access to the sum variable. A more efficient technique is to allow each thread to collect partial sums of the sum variable and then to add them together after the threaded loop. Figure 6.14 shows how this is done.

Still another more efficient method of implementing the code is to use an OpenMP reduction clause as detailed in Figure 6.15. The reduction clause is semantically the same as the partial sums version in Figure 6.14 except the OpenMP run-time manages collection of the partial sums for you.

The use of synchronization is a general must-have for multi-threaded development and learning techniques for reducing the amount of necessary synchronization helps maximize performance.

Thread Stall

Another method of ensuring shared resources are correctly accessed is through a lock. In this case, a thread will lock a specific resource while it is using that resource, which also denies access to other threads. Two common thread-related errors can occur when using locks. The first is a thread stall. This happens when you have one thread that has locked a certain resource and then moves on to other work in the program without first releasing the lock. When a second thread tries to access that resource it is forced to wait for an infinite amount of time, causing a stall. A developer should ensure that threads release their acquired locks before continuing through the program.

Deadlock

The third common thread-related error is deadlock. A deadlock is similar to a stall, but occurs most frequently when using a locking hierarchy. If, for example, thread one locks variable A and then wants to lock variable B while thread two is simultaneously locking variable B and then trying to lock variable A, the threads are going to deadlock. Both are trying to access a variable that the other has locked. In general, you should avoid complex locking hierarchies if possible as well as insuring that locks are acquired and released in the same order. Recoding thread two so that it locks variables in the same order, A first and then B, would help prevent the deadlock in the example.

6.2.3.3 Logging

The data race example illustrated in Figure 6.4 shows that the order that threads execute is important for correct execution. A typical debugger cannot display the history of execution leading up to the current stop point in a debug session. Therefore a means of logging trace information of program behavior can help during debugging. One technique to accomplish logging is for the programmer to create a buffer area where the application emits messages containing thread identification and status. A programmer should consider the amount of logging to enable in a production build versus a *debug build* of the application.

6.2.3.4 Finding Thread-related Bugs

Now that you understand the types of thread-related bugs that can exist in code, it is time to find them. Tools to assist finding thread-related bugs were discussed in Chapter 4. These tools use binary instrumentation and modify an application to test certain conditions on the application's memory references during execution time. The tools place substantial execution time and system impact overhead on the targeted application. For example, a sample application can execute 20 times slower when undergoing the analysis. The following recommendations help maximize the finding of thread-related bugs using the tools:

- Use a debug and non-optimized build of your application

- Ensure threaded regions are executed in your benchmark

- Reduce the memory image size

- Reduce the workload of the benchmark

- Limit instrumentation

It is recommended, regardless of how instrumentation is done, that a debug build be used that includes symbols and line numbers, no optimization is used, and the binary be relocatable (for binary instrumentation). Keeping debug symbols and line numbers will give the thread verification tool the chance to point directly to source lines that have possible problems; turning off all optimization will keep the application code closest to the original source order (and if there is a threading error with optimization, but no problem without optimization, the problem is more likely in the compiler and not your threading).

The second recommendation is crucial – make sure the code paths containing your multi-threaded code are executed by the benchmark. The tools cannot analyze for thread issues unless the actual threaded code paths are executed. The last three recommendations run counter to the previous in that while you want to ensure your threaded regions execute, you also want to minimize the memory use and execution time because of the binary instrumentation overhead. The third recommendation is to make sure your application clearly signals when it has executed correctly. Since the application execution time will be longer when using the thread verification tool, it may be the case that you start the test run and leave the system unattended during execution. A hypothetical example involves a haphazard coding practice that allocates memory as detailed in Figure 6.16.

```
X = malloc(size);
   if (X==NULL) exit(0);
```

Figure 6.16: Typical memory allocation code

One positive aspect of the code is that the return value from malloc is checked and handles cases where *X* is NULL such as when no memory is available. The negative aspect is that the application exits with no visible message to the user regarding the cause. The issue with the code is that the binary instrumentation associated with thread verification tools adds to the memory usage of the instrumented application. The addition may cause the user application to run out of memory and because the application execution time is exaggerated, you may be executing the application in a batch mode and not explicitly watching program behavior. As a result, you may be lead to believe the application worked and the thread tools produced diagnostics that indicate no issue while what really happened is your application never exercised the relevant threaded code.

The fourth recommendation is reduce the workload of the application under test. If the application was parallelized via data decomposition, look for techniques of reducing the total iteration space exercised when running the verification tool.

Finally, the last tip is to limit the instrumentation. If the threaded region is confined to a small area, there is no point having the tool instrument the entire application. It may be possible to instrument just a small portion of the relevant application and thus cut down on the execution time overhead of binary instrumentation.

6.2.4 Tune

The last phase of the TDC is Tuning, where application performance is examined and potential thread-related performance limiters are addressed. The types of questions asked during this phase include:

- How much of the application is running in parallel?
- Is the work evenly distributed between threads?
- What is the impact of synchronization between threads on execution time?
- Is memory effectively used and shared between the threads and processor cores?

The amount of the application which runs in parallel is important because the speedup attributed to parallelization is limited by the serial portion. We learned this important fact in our discussion of Amdahl's Law earlier in the chapter. Using a system monitoring tool such as Perfmon or the mpstat command helps provide an estimate of this amount; the developer can observe how much and when the application is taking advantage of the

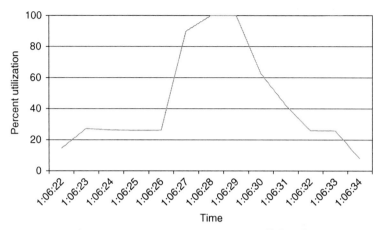

Figure 6.17: Mpstat output on parallel code

processor cores. Figure 6.17 is a graph of the output of mpstat executing on a system with four processor cores and executing a multi-threaded program. An easy characterization is that during the first third of program execution time only one processor core is utilized. During the second third of program execution time, all processor cores are utilized. During the last third of program execution time, one processor core is utilized. One conclusion is that if more performance from parallelism is desired, the portion of code executing during the first and last third should be analyzed.

Even distribution of work to the threads helps ensure optimal processor core utilization. If the workload is unbalanced, there may be some threads that are sitting idle with no work to accomplish while other threads are executing. An improved distribution of work can help ensure all processor cores are used as effectively as possible. Take for example, the type of profile returned by Intel® Thread Profiler. Based upon the profile, the developer can determine if a different balancing of the workload can utilize the processor cores more effectively. The classic example is a triangular matrix operation where a sharing scheme based upon division of loop iterations may disadvantage threads with larger data ranges. Figure 6.18 is the output of Intel® Thread Profiler after measuring the thread balance on a program to compute prime numbers [8]. The algorithm used to find prime numbers in the application splits the range of numbers to check based upon the number of threads and assigns the work such that each thread checks a monotonically increasing range. In other words, if two threads were assigned to check for primes from 1 to *N*,

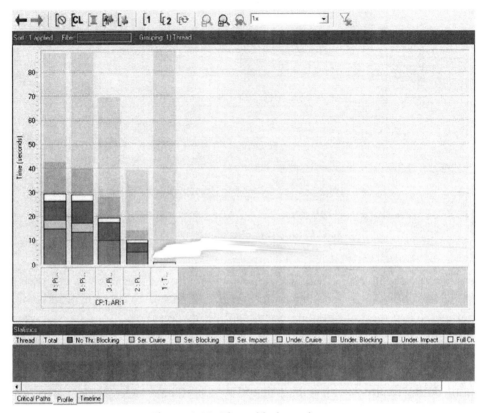

Figure 6.18: Thread balance issue

thread one would check from 1 to *N*/2 and thread two would check from *N*/2 + 1 to *N*.
The result is a load imbalance because threads that check a range that is less than the
range that another thread checks have much less work to do. The solution to this problem
is to select a different workload distribution. In the case of the prime number code the
solution is to distribute the work in chunks of 100 numbers so each thread checks primes
on numbers of similar magnitude.

6.2.4.1 Synchronization

Synchronization is a necessary aspect of parallelism that impacts performance. The
overhead attributed to synchronization depends upon the mechanism used to implement

it. The utility mpstat and the Intel® Thread Profiler are two tools that help in determining synchronization overhead and lock contention. It is admittedly difficult to determine synchronization overhead using mpstat; however, a clue that is indicative of a problem is observing a large proportion of kernel time and a large number of context switches per second on a multi-core system when your application is executing. Intel Thread Profiler instruments thread API calls and reports on time spent inside of synchronization routines. If synchronization time was determined to be excessive, you could analyze your code to see how to simplify or lower the performance effects of the synchronization. Lowering the performance impact of synchronization is accomplished by restructuring your algorithm to reduce the frequency of the sharing. For example, it may be possible for each thread to compute on private copies of a value and then synchronize the value at the end of the computation similar to how the reduction clause works in OpenMP.

Synchronization is also accomplished by adding mutual exclusion around data items and regions of code. One recommendation when employing synchronization is to understand the advantages and disadvantages of your synchronization method. For example, Windows threads offer critical sections and mutexes for synchronization. A mutex is a kernel object that has the advantage of being usable between processes with the disadvantage of executing more slowly than a critical section.

6.2.4.2 Memory Hierarchy

The discussion of techniques to reduce synchronization leads to the last area which is effective use of the memory hierarchy. Similarly to how multi-core processors enable new classes of bugs in your application, multi-core processors present additional challenges with regard to memory and cache optimization. Optimizing the cache for multi-threaded execution can substantially improve performance. Several techniques for optimizing for the caches of multi-core processors include:

- Align data on hardware boundaries (avoid cache splitting)

- Use hardware/software prefetch properly

- Use *spatial locality* and *temporal locality* to your advantage

- Use processor affinity

- Minimize false sharing

The first two approaches, align data on hardware boundaries and prefetch, are similar to serial optimization performance techniques and are not detailed here. With regards to taking advantage of spatial and temporal locality, one common multi-core technique involves delaying the handoff of shared values from one thread to another by increasing per thread computation. This allows the cache hierarchy to evict these shared values from cache to a higher level that is shared between threads.

Processor affinity is the explicit binding of threads to processor cores. Modern OSes make no guarantees about which processor core will run which thread and whether or not the same processor core runs the same thread from context switch to context switch. Affinity instructs the OS to pin a particular thread to a specific processor core. A simple case where affinity provides benefit is where a producer thread and consumer thread residing on one processor are able to share data in a higher level cache compared to the two threads residing on different processors and synchronizing data through main memory.

The final issue in this section is false sharing. If you recall, false sharing occurs when two threads manipulate data that lie on the same cache line. One way to detect false sharing is to sample on L2 cache misses using a profile tool that supports hardware EBS. If this event occurs frequently in a threaded program, it is likely that false sharing is at fault.

Chapter Summary

The TDC is a process to find where to thread an application and how to design, debug, and tune the implementation. The analysis step focuses on how to find the most important regions of your applications to thread. The design and implement step involves the actual coding work of the threads and is aided by knowing techniques to ease the introduction of threads into your application. The debug step includes finding and fixing both non-thread-related bugs and thread-related bugs. Fourth, the tuning step considers how to diagnose thread performance issues and methods to improve performance. One final point is that although the TDC consists of four steps, the application of this process is a cycle that does not necessarily conclude after step four. Instead the process of analysis, design and implement, debug, and tune may be repeated several times until the application performance requirements are met. In some cases, a code change at step four to improve performance may introduce stability bugs and require debug work at step three. Significant stability changes may require a reanalysis of the application as the hotspots may have changed. Consider the TDC as a process for developing threaded code that is a cycle and can jump from step to step as most appropriate to meet your end goal.

Related Reading

This chapter has the challenging task of attempting to cover the TDC to a level of detail great enough for the reader to understand and apply to their embedded project. At the same time, the details cannot be so great as to apply to only a relative few readers. As such, it makes sense to mention related books where a reader can find greater details on particular topics.

The first book to recommend is "Programming with Hyper-Threading Technology" [9] by Andrew Binstock and Richard Gerber. This book was written before the advent of multi-core processors but provides good fundamental information on multi-threading. A related book, "Multi-Core Programming" [10] by Shameem Akhter and Jason Roberts provides coverage of SMP programming on Intel Architecture and greater detail on thread technology like Win32, Pthreads, and OpenMP. Even more specifics on Pthreads are found in "Programming with POSIX® Threads" [11] by David R. Butenhof. The final reference is to "Patterns for Parallel Programming" [12] by Timothy G. Mattson, Beverly A. Sanders, and Berna L. Massingill, which details the mechanics of finding and exploiting concurrency in your application.

References

[1]   SETI@home, http://setiathome.berkeley.edu/

[2]   J. L. Hennessy and D. A. Patterson, *Computer Architecture – A Quantitative Approach*. San Francisco, CA, Morgan Kaufmann, 2003.

[3]   D. Levinthal, Cycle Accounting Analysis on Intel® Core™ 2 Processors, http://assets.devx.com/goparallel/18027.pdf

[4]   SPEC 2000, http://www.spec.orgwww.spec.org

[5]   G. A. Carpenter and S. Grossberg, ART 2: Stable self-organization of pattern recognition codes for analog input patterns. *Applied Optics*, 26, 4919–4930, 1987.

[6]   SPEC OMP, http://www.spec.org/omp/

[7]   V. Aslot, M. Domeika, R. Eigenmann, G. Gaertner, W. B. Jones and
      B. Parady, SPEComp: A New Benchmark Suite for Measuring Parallel Computer
      Performance. *Workshop on OpenMP Applications and Tools*, pp. 1–10, July 2001,
      http://citeseer.ist.psu.edu/aslot01specomp.html

[8]   C. Breshears, *Intel® Threading Tools and OpenMP**, http://www3.intel.com/cd/
      ids/developer/asmo-na/eng/292206.htm?page=1

[9]   A. Binstock and R. Gerber, *Programming with Hyper-Threading Technology*. Intel
      Press: Hillsboro, OR, 2004.

[10] S. Akhter and J. Roberts, *Multi-Core Programming*. Hillsboro, OR: Intel Press,
      2006.

[11] D. R. Butenhof, *Programming with POSIX(R) Threads*. Reading, MA: Addison-
      Wesley Professional, 1997.

[12] T. G. Mattson, B. A. Sanders, and B. L. Massingill, *Patterns for Parallel
      Programming*. Reading, MA: Addison-Wesley Professional, 2004.

## CHAPTER 7

Case Study: Data Decomposition

Key Points

- Performance analysis of a GUI application must be performed carefully, oftentimes requiring use of special steps and tools' features.

- Determining the location to insert threading is based upon a number of factors including frequency of execution, source code modifications, and threading overhead.

- When implementing threading, manually inspect the code to determine variables that need to be shared and private. Use the thread verification tool for difficult to find threading issues.

The threading development cycle (TDC) provides a useful step-by-step process that can be applied to a threading project. This chapter employs the process on a real application where the concurrent work is found via data decomposition. Each step in the process is documented with details on the thought process, tools usage, and results. I intend this case study to relay the actual steps taken and observations made during the optimization process. Encountered issues are described as they occurred in the process and solutions then explained. I avoided rewriting the steps in the case study to accommodate information discovered later in the process in the attempt to produce a neatly packaged study; real thread projects rarely fit a prescribed process that has no flexibility. The expectation is that you, the reader, could reproduce the steps and results. The end result is a speedup of 2.267 times the performance of the original application executing on a four processor core system.

7.1 A Medical Imaging Data Examiner

A Medical Imaging Data Examiner (AMIDE) [1] is an open source tool that provides visualization of medical data. The tool takes as input a variety of raw image data files and supports viewing of the data and a number of visual processing transformations. One key function, volumetric rendering, translates three-dimensional data to two dimensions for viewing in real time. To accomplish the volumetric transformations, AMIDE relies upon a software library called VolPack [2]. Both AMIDE and VolPack are written in C and total approximately 250,000 lines of source code. This case study focuses on improving the performance of the volume rendering feature of AMIDE.

The input data set is comprised of two different types of scans of a mouse in several positions [3]. The volumetric rendering is performed by mapping the visibility of three-dimensional data onto a two-dimensional plane when viewed from a particular angle using traditional *ray tracing* techniques. VolPack employs a shear warp transformation on the data set that allows efficient object-order transformation [4]. Figure 7.1 is a screenshot of the volumetric rendering window under AMIDE employing the input data set.

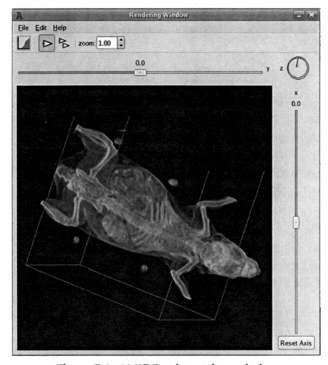

Figure 7.1: AMIDE volumetric rendering

7.1.1 Build Procedure

The basic build procedure for AMIDE is documented in Table 7.1. The steps consist of downloading AMIDE and VolPack, setting up the build directories, building VolPack, and building AMIDE.

Removing the current build of AMIDE and VolPack in preparation for a new build is accomplished by typing:

```
make distclean
```

in both the AMIDE-0.8.19 and VolPack-1.0c4 directories. The optimization process employs different CFLAGS and LDFLAGS settings during the configuration and build steps of AMIDE and VolPack. The build procedure for these optimized builds is similar to the steps documented in Table 7.1 except for the `configure` and `make` commands; these differences will be detailed at each step.

7.1.2 Analysis

The analysis of AMIDE is preparation for the eventual parallel optimization effort, which includes serial optimization, and a study of the application to determine the best locations to thread.

7.1.2.1 Serial Optimization

Before threading the application, two compilers and several compiler optimizations are employed on the serial version of the code. This serial optimization is a prerequisite for parallel optimization and serves a dual purpose. First, the serial optimization provides a baseline of performance for the parallel optimization effort; second, the optimization effort provides familiarity with the application. The serial optimization steps and results are summarized in Table 7.2.

Characterization of the application reveals the following:

- Heavy floating point computation – the application is calculating three-dimensional rotations of an object and the pixels that are visible at a particular angle. The data structures that deal with the input data and calculations are predominantly single-precision floating point.

Table 7.1: AMIDE build procedure

Download and untar files[a]	`mkdir <build directory>`
	Download `amide-0.8.19.tgz` [5]
	Download `volpack-1.0c4.tgz` [6]
	Download `m2862-small.xif.gz`
	Untar and unzip the files
	`tar -zxvf amide-0.8.19.tgz`
	`tar -zxvf volpack-1.0c4.tgz`
	`gunzip m2862-small.xif.gz`
	`cp m2862-small.xif ./amide-0.8.19`

Configure and build VolPack	`cd volpack-1.0c4`
	`./configure`
	`make`

Setup VolPack include and library directories	`mkdir include`
	`ln -s ../src/volpack.h include/volpack.h`
	`mkdir lib`
	`ln -s ../src/.libs/libvolpack.a lib/libvolpack.a`
	`ln -s ../src/.libs/libvolpack.la lib/libvolpack.la`
	`ln -s ../src/.libs/libvolpack.so lib/libvolpack.so`
	`ln -s ../src/.libs/libvolpack.so.1 lib/libvolpack.so.1`
	`ln -s ../src/.libs/libvolpack.so.1.0.4 lib/libvolpack.so.1.0.4`
	`mkdir lib/.libs`
	`ln -s ../src/.libs/libvolpack.a lib/.libs/libvolpack.a`
	`ln -s ../src/.libs/libvolpack.la lib/.libs/libvolpack.la`
	(Continued)

Table 7.1: (Continued)

	`ln -s ../../src/.libs/libvolpack.so lib/.libs/libvolpack.so`
	
	`ln -s ../../src/.libs/libvolpack.so.1 lib/.libs/libvolpack.so.1`
	
	`ln -s ../../src/.libs/libvolpack.so.1.0.4 lib/.libs/libvolpack.so.1.0.4`

Configure and build AMIDE	`cd ../amide-0.8.19`
	
	`./configure LDFLAGS="-L/<build directory>/volpack-1.0c4/lib`
	
	`make CFLAGS=-I/<build directory>/volpack-1.0c4/include`

Execute AMIDE	`cd src`
	
	`./amide ../m2862-small.xif`

Perform volume rendering	Select *View* menu item and *Volume Rendering* Click on *Execute*

[a] Both AMIDE and VolPack are under development so version numbers may change. These instructions should still work for the updates; however, some modifications may be required.

Table 7.2: AMIDE serial optimization summary

Serial optimization step	Results
Characterize the application	Single and double precision floating point intensive, loop intensive processing
Prioritize compiler optimization	Automatic vectorization, data placement optimizations
Select benchmark	Volume rendering along *X*- and *Y*-axis rotations
Evaluate performance of compiler optimizations	Achieved a 54% improvement over the baseline optimization settings

- Multiple fairly tight loop nests – the application consists of nested loops that iterate over the image data. One could assume the processing will be stream oriented, sweeping across a wide range of data, and performing fairly constrained computations.

The compiler optimizations to apply should target floating point loop optimizations. Therefore, automatic vectorization and loop optimizations are prioritized high. Profile-guided optimization will be used with the expectation of providing the automatic vectorization phase with better loop bounds and also benefiting application code size. A benchmark will be developed to isolate the volume rendering functionality and will focus on performing a series of rotations along the *X*-axis and the *Y*-axis that is automated and reproducible from run to run. Using the compiler optimization settings, a 54% execution time improvement was obtained over the baseline optimization settings. The next few sections provide further detail on each phase of serial optimization.

7.1.2.2 Benchmark

Control in AMIDE is accomplished via a GUI with menu options providing access to different functions. The particular function targeted for optimization in this case study is the volumetric rendering which is accessed by selecting the "View" menu and clicking on "Volume Rendering." A window appears and clicking on the "Execute" button starts loading the image. Once the image is loaded, sliders can be adjusted to position the viewing angle of the object at any number of locations. For the purposes of optimization, an interactive application that relies upon constant user input is not easily reproducible from execution to execution. Therefore, the source code is modified by removing the need for user input during the critical volume rendering portion of the benchmark. The source code is modified to pass through 20 rotations on the *X*-axis, 20 rotations on the *Y*-axis, and output timing information. The timing information is low overhead in nature and records times at the start and completion of each rotation and the entire set of rotations. A sample of timing output from an execution of the benchmark appears in Figure 7.2. The key output is the time denoted after "Rendering all objects took" which indicates the time from start to finish of the 40 rotations. In general, if a fair amount of modification is required to create the benchmark, it is recommended to validate the several steps done in the analysis such as creating the execution time profile and testing different optimizations against the original version of the code.

7.1.2.3 Serial Optimization Results

The serial optimization phase employs multiple compiler optimization levels to improve the performance of the application. Compiler, optimization settings, and timing results are summarized in Table 7.3. Timing results are calculated based upon the median of three executions of the 40 rendering positions as described previously. The system where the

```
Stage 1 Render 15
######## Rendering objects took 4.344 (s) #########
Stage 1 Render 16
######## Rendering objects took 4.375 (s) #########
Stage 1 Render 17
######## Rendering objects took 3.820 (s) #########
Stage 1 Render 18
######## Rendering objects took 3.817 (s) #########
Stage 1 Render 19
######## Rendering objects took 3.867 (s) #########
######## Rendering all objects took 196.565 (s) #####
```

Figure 7.2: Sample benchmark output

Table 7.3: AMIDE serial optimization settings and results

Compiler	Optimization setting	Time result (in seconds)	Improve (%)
gcc 3.4.3 (default)	VolPack: `-O2 -g` AMIDE: `-g -O6`	196.893	0
gcc 3.4.3	`-O3 -mtune=pentium-m`	105.567	46
icc 9.1	`-O2`	93.785	52
icc 9.1	`-O3`	93. 77	52
icc 10.0	`-O3 -xP`	90.68	54
icc 10.0	`-O3 -ip -xP`	90.78	54
icc 10.0	`-O3 -ipo -xP`	90.543	54
icc 10.0	`-O3 -ipo -xP` `-prof-use`	90.157	54

tests were executed is an Intel Core Duo processor with a clock speed of 2.0 GHz, 1 GB RAM, and executing the Red Hat Enterprise Linux 4 operating system.

Optimization settings were added as options to the `configure` command and/or the `make` command used when building VolPack and AMIDE. Three different compilers were employed, gcc 3.4.3 and the Intel C++ Compiler versions 9.1 and 10.0. GNU gcc 3.4.3 is the default compiler on the operating system. A summary of the commands to build using icc 10.0 and the options "`-O3 -ipo -xP`" is detailed in Table 7.4.

A couple of additions to the command lines were required to enable the advanced optimization build with icc. First, LD and AR were set equal to `xild` and `xiar`, respectively. The programs, `xild` and `xiar`, supplement the standard `ld` and `ar` on the

Table 7.4: AMIDE ICC command line

Configure and build VolPack	`./configure CC=icc CXX=icc LD=xild AR=xiar CFLAGS="-O3 -ipo --xP" LDFLAGS="-O3 -ipo -xP"`
	`make`
Configure and build AMIDE	`./configure CC=icc CXX=icc LD=xild LDFLAGS="-L/<build directory>/ volpack-1.0c4/lib -O3 -ipo -xP -lsvml"`
	`make CFLAGS="-I/<build directory>/ volpack-1.0c4/include -O3 -ipo -xP"`

system to enable the build with interprocedural optimization (`-ipo`). Without this option, a series of undefined symbol errors would be encountered during archive creation. The environment variable, LDFLAGS, specifically lists the compiler optimization settings such as `-O3` and `-xP` because interprocedural optimization occurs at what is traditionally link time. In addition, the `-xP` option enables automatic vectorization and requires an explicit link against the short vector math library (`svml`) and thus the `-lsvml` option is added.

The last optimization settings included the use of profile-guided optimization. Table 7.5 lists the specific configure and make command lines for building and using profile-guided optimization. The `-prof-dir` option is employed to specify a location where the profile data files are dumped during the execution of AMIDE. In general it is a good idea to place your profiles in one location. The profile generation phase employs the benchmark as the workload because the execution time increase from profiling is not substantial. In a production environment, you are strongly encouraged to use a mix of different input data sets for the profile generation phase because typically one data set is not inclusive of all the possible inputs your customers may provide. In addition, it is possible for the compiler to over optimize for the limited set of inputs at the expense of the breadth of customer input.

One concern when applying profile-guided optimization is verification that the profile contained useful data. One method of verification is to scan the compilation output for the string "profile info" and then determine if some of the routines used profile information. Table 7.6 shows a subset of the results of searching the compilation output for "profile info." The output shows the file, `vp_check.c`, contained 10 routines, 6 routines that

Table 7.5: AMIDE serial optimization-profile guided

Configure and build VolPack: Profile generation	`./configure CC=icc CXX=icc CFLAGS= "-prof-gen -prof-dir <build directory>/pgo"` `make`
Configure and build AMIDE: Profile generation	`./configure CC=icc CXX=icc LDFLAGS= "-L/<build directory>/volpack-1.0c4/lib "` `make CFLAGS="-I/<build directory>/volpack-1.0c4/include --prof-gen -prof-d ir <build directory>/pgo"`
Configure and build VolPack: Profile use	`./configure CC=icc CXX=icc LD=xild AR=xiar CFLAGS="-O3 -ipo -xO -prof-use -prof-dir <build directory>/pgo" LDFLAGS=" -O3 -ipo -xO -prof-use -prof-dir <build directory>/pgo"` `make`
Configure and build AMIDE: Profile use	`./configure CC=icc CXX=icc LD=xild LDFLAGS="-L/<build directory>/volpack-1.0c4/lib -O3 -ipo -xO -lsvml -prof-use -prof-dir <build directory>/pgo"` `make CFLAGS="-I/<build directory>/volpack-1.0c4/include -O3 -ipo -xO -prof-use -prof-dir <build directory>/pgo"`

had profile information, and 4 routines that did not have profile information. This simple search provides a level of confidence that the profile generated during the execution of the -prof-gen compilation is being found and associated with routines.

A second technique for verifying profile-guided optimization information with the Intel compiler is to execute the profmerge[1] command with the -dump option. One

---

[1] The executable, profmerge is located in the Intel compiler bin directory.

Table 7.6: AMIDE profile-guided verification

`$ grep "profile info:" volpack-compiler-output.txt`
`vp_check.c: warning #11503: Total routines 10, routines w/o profile info: 4`
`vp_context.c: warning #11503: Total routines 40, routines w/o profile info: 27`
`vp_extract.c: warning #11503: Total routines 11, routines w/o profile info: 11`
`vp_file.c: warning #11503: Total routines 15, routines w/o profile info: 15`
`vp_linalg.c: warning #11503: Total routines 11, routines w/o profile info: 7`
`vp_octree.c: warning #11503: Total routines 16, routines w/o profile info: 15`
`vp_renderB.c: warning #11503: Total routines 9, routines w/o profile info: 9`

of the first values output is labeled "MISMATCH OCCURRED:" and is equal either to "F" or "T". An "F" indicates that the profile information matches the source files (no modifications to the source file occurred after the profile was generated).

Performing serial optimization, an improvement of 54% was obtained. The difference between optimization using icc with $-O2$ and with the advanced optimizations was very minimal. If further performance via serial optimization was desired, the next step would be to find the most frequently executed regions of code and focus optimization on those areas. For the case study, the serial optimization results employing the "$-O3$ $-xP$" options serve as the baseline to compare against the parallel version.

7.1.2.4 Execution Time Profile

GNU gprof was used to obtain the execution time profile. Enabling the gprof build, required the command lines detailed in Table 7.7 to be modified in building VolPack and AMIDE.

The executable, `.libs/lt-amide`, is specified instead of the script, AMIDE. This change over previous executions informs gprof of the executable of interest; however, AMIDE will need to be executed beforehand. The initial execution of the script creates

Table 7.7: AMIDE build with GPROF

Configure and build VolPack	`./configure CFLAGS=-pg LDFLAGS=-pg`
	
	`make`

Configure and build AMIDE	`./configure LDFLAGS="-L/<build directory>/`
	`volpack-1.0c4/lib -pg -static"`
	
	`make CFLAGS="-I/<build directory>/volpack-`
	`1.0c4/include -pg -static"`

Generate gprof output	Execute amide and exit (creates `.libs/lt-amide`).
	
	`cd src`
	
	`./amide`
	
	Execute AMIDE and perform Volume Rendering
	
	`gprof .libs/lt-amide ../m2862-small.xif`

```
Flat profile:

Each sample counts as 0.01 seconds.
  %   cumulative   self              self     total
 time   seconds   seconds    calls   s/call   s/call  name
92.55   6752.68   6752.68   120568    0.06     0.06   VPCompAR11B
 2.70   6949.33    196.65      527    0.37     0.40   amitk_data_set_SSHORT_2D_SCALING_get_slice
 1.61   7066.91    117.58      241    0.49    28.89   image_from_renderings
 1.00   7140.02     73.11        2   36.56   161.49   rendering_load_object
```

Figure 7.3: AMIDE Gprof execution time profile

`.libs/lt-amide`. Figure 7.3 details the execution time profile of the AMIDE benchmark. The top region of interest is the function `VPCompAR11B()` which accounts for 92.55% of the execution time recorded by the profile. One critical modification in creating the execution time profile is due to a limitation of gprof. By default, gprof does not profile shared libraries even if the shared library was compiled using the `-pg` option. There are three solutions to this issue. If you do not employ one of these solutions, your generated profile may not have accurate information and as a result you may miss the critical functions that would provide the most benefit to you from optimization. In the case of AMIDE and without the fix for shared libraries, `amitk_data_set_SSHORT_2D_SCALING_get_slice()` is incorrectly labeled as the most time-consuming function with no mention of `VPCompAR11B()`.

The three options for generating a profile that obtains information across shared libraries are:

1. Set the LD_PROFILE environment variable to the path of the shared library.

2. Link against the library statically.

3. Use a profiling tool that measures across the entire system.

To use the first technique, after execution, execute gprof with the name of the shared library and the corresponding gprof output file as command line arguments.

For this case study, the second option was employed which is to link statically against libvolpack.a. The option -static was added to the CFLAGS and LDFLAGS in the build of AMIDE. The third option is to employ a tool such as the Intel® VTune™ Analyzer, which has the ability to create a profile and record data across shared libraries.

For verification of the gprof profile, an event-based sampling (EBS) profile is generated using VTune™ Analyzer. Specifically, version 9.0.0 of VTune™ Analyzer and the "First Use" tuning activity was selected when creating the profile. The following settings were applied and are depicted in Figure 7.4:

- Application to launch – <Build directory>/amide-0.8.19/src/exe/lt-amide

- Application arguments – m2862-small.xif

- Working directory – <Build directory>/amide-0.8.19

Please note the small addition of an exe directory which is a symbolic link to the .libs directory:

```
ln -s <build directory>/amide-0.8.19/src/.libs <build
directory>/amide-0.8.19/src/exe
```

VTune™ Analyzer is unable to change directories to a directory beginning with ".". In addition, to execute the GUI of VTune™ Analyzer required the following environment command:

```
export OTHER_JVM_BINDIR=/opt/intel/eclipsepackage/3.2.1/
jrockit-R27.2.0-jrel.5.0_10/bin
```

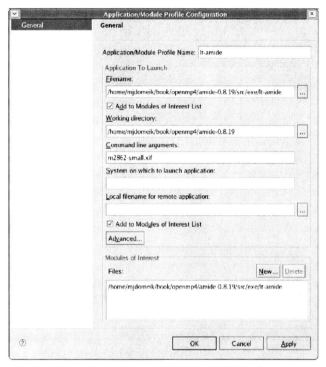

Figure 7.4: VTune Analyzer configuration

The profile results indicate that 89.78% of the profiling samples occurred in
`VPCompAR11B()` which validates the results of the gprof profile. One nice property of the
EBS profile is that the events are correlated with source code enabling a finer view of the
execution behavior. Figure 7.5 is a screenshot of the profile and highlights some of the key
contributors to the execution time. Highlighting line 4955 to line 5055 of `vp_compAR11B.c`
indicates 87.6% of the samples were collected in that region. This region is the critical
portion of the execution time and will be further scrutinized when loops are analyzed.

An additional use for the EBS profile is to estimate potential speedup from threading. Before
doing so, the code must be modified to enable VTune™ Analyzer to focus on the specific
code targeted for optimization. One challenge when using VTune™ Analyzer on a GUI
application such as AMIDE is that user interaction is required. The application executes and
waits on user input, which by nature forces a degree of variability in the measurements. This
is bad. Second, the EBS profile may include events from code outside of the targeted code.

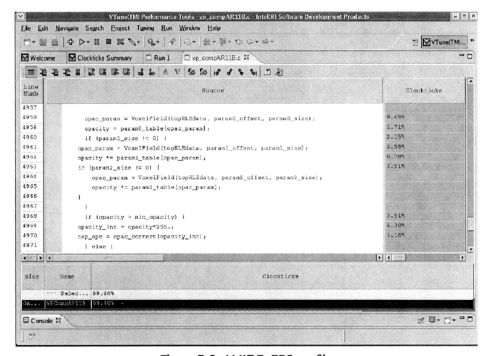

Figure 7.5: AMIDE: EBS profile

Therefore, the metrics may be skewed by events recorded during application initialization, shutdown, and times when the application is waiting on user input. In order to control for variability and focus the profile on the targeted code, a technique of pausing and resuming the profiling is employed. VTune™ Analyzer supplies libVtuneApi.so, a library that contains functions VTPause() and VTResume, which pause and resume profiling, respectively. Table 7.8 summarizes the modifications necessary to enable this functionality.

The sources are modified by adding a call to VTPause() at the start of main(). As soon as control reaches this point during the profiling run, collection of events is suspended. A call to VTResume() is added right before the start of the benchmark loop in ui_render.c/canvas_event_cb() which starts the collection of events. A call to VTPause() is added after the benchmark loop to once again suspend collection of events. LDFLAGS for building AMIDE is modified to include the directory that contains libVtuneApi.so and explicitly specifies use of libVtuneAPI.so. CFLAGS for building AMIDE is modified to provide the include path for VtuneApi.h. Finally, the

Table 7.8: AMIDE-VTPause() and VTResume

Modify sources	Add VTPause() at the start of main() in amide.c. Add:
	#include <VtuneApi.h> at the start of amide.c
	Add VTResume() right before the benchmark loop in ui_render.c/canvas_event_cb() Add:
	#include <VtuneApi.h> at the beginning of the file.
	Add VTPause() after the benchmark loop in canvas_event_cb().
Configure and build AMIDE	./configure LDFLAGS="-L/<build directory>/volpack-1.0c4/lib
	-L/opt/intel/vtune/analyzer/bin -lVtuneAPI"
	make CFLAGS="-I/<build directory>/volpack-1.0c4/include -I/opt/intel/vtune/analyzer/include"
Set environment	export LD_LIBRARY_PATH=$LD_LIBRARY_PATH; /opt/intel/vtune/analyzer/bin"

LD_LIBRARY_PATH is appended to enable the loader to locate libVtuneAPI.so during application execution.

Once the preceding build modifications are made, an EBS profile is ready to be collected. In order to ease analysis of the collected data, a uniprocessor version of the Linux kernel is booted even though the execution system contains a dual-core processor.

Figure 7.6 is a screenshot of the EBS profile displaying the most active functions after employing the preceding VTPause() and VTResume() modifications. Figure 7.7 is the result of clicking on the libvolpack.so.1.0.4 link and provides a report on the most active functions inside the library.

The EBS profile indicates 94.18% of the samples occurred inside of VolPack and almost all of the samples occurring inside of VolPack were in VPcompAR11B(). Amdahl's Law is applied to estimate the speedup limits of the threading which appear in Table 7.9.

In these calculations, we assume that the functions targeted for multi-threading will exhibit linear scaling with regards to the number of processors. Accordingly, the time of the parallel portion for the four processor core case is 0.235 which is 0.94/4. Speedup

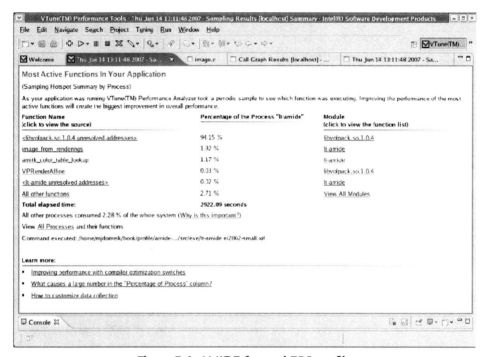

Figure 7.6: AMIDE focused EBS profile

is computed by dividing the baseline time results by the time results for a particular execution. In this case, the baseline is the one processor core time. The theoretical limit on speedup in the four processor core case is 3.39 times.

7.1.2.5 Collect a Call Graph Profile

The previous execution of gprof on the benchmark also produced a call graph profile of AMIDE which is detailed in Figure 7.8.

The call graph reveals that VPRenderAffine() is the primary caller of VPCompAR11B() and examination of the source code reveals the call occurs through an indirect function call referenced through the variable, composite_func. VPRenderAffine() and VPCompAR11B() are located in the files, volpack-1.0c4/src/vp_renderA.c and volpack-1.0c4/src/vp_compAR11B.c, respectively. The profile recorded 482 invocations of VPRenderAffine() and 120568 invocations of VPCompAR11B().

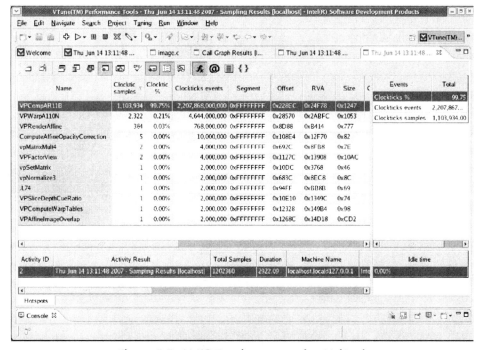

Figure 7.7: AMIDE estimate speedup VolPack

Table 7.9: AMIDE theoretical speedup

Processor cores	Serial portion	Parallel portion	Total time	Speedup
1	0.06	0.940	1.000	1.00
2	0.06	0.470	0.530	1.89
4	0.06	0.235	0.295	3.39

```
index % time    self  children    called       name
------------------------------------------------------
               0.48  6775.32     482/482        vpRenderRawVolume [5]
[6]    92.9    0.48  6775.32     482            VPRenderAffine [6]
              6752.68   0.00   120568/120568    VPCompAR11B [7]
              22.61     0.03      482/482        VPWarpA110N [20]
               0.00     0.00      482/482        VPSliceDepthCueRatio [187]
------------------------------------------------------
              6752.68   0.00   120568/120568    VPRenderAffine [6]
[7]    92.5  6752.68   0.00   120568          VPCompAR11B [7]
```

Figure 7.8: AMIDE call graph

7.1.2.6 Flow Chart Hotspots

On many performance optimization projects, you may not be the original author of the code and therefore may not understand the details of the code including, for example, the algorithms being used, the data layout, and access patterns. Creating a flow chart of the hotspots helps provide this information to you. In addition, the flow chart serves as a basis for the eventual source modifications for threading. Figure 7.9 represents a flow chart for the hotspots determined by the execution time profile and call graph profile obtained in the previous steps.

The processing in VPRenderAffine() takes as input volumetric data consisting of voxels. Voxels are a shortened combination of volume and pixel and represent volume in three dimensions similar to how a pixel represents a point in two dimensions.

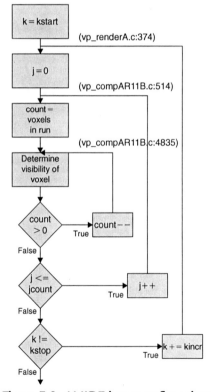

Figure 7.9: AMIDE hot spot flow chart

In VPRenderAffine(), the voxels are divided into slices and a call is made to VPCompAR11B() for further processing. Each slice is divided into scanlines and bounded to determine the dimensions of the slice that could be in the final image. Each scanline of the image is then processed on a pixel-by-pixel basis to determine if the pixel should be in the composite two-dimensional image and the pixel's opacity and color.

7.1.2.7 Classify Most Frequently Executed Loops

The most frequently executed loops are determined by analyzing the flow chart information collected previously. The flow chart indicates processing is centered around three nested loops that iterate across the range of slices, scanlines, and pixels, respectively.

Table 7.10 is a classification summary of the most frequently executed loops. The analysis reveals that the primary loops are in the functions VPRenderAffine() and VPCompAR11B(). The loop at vp_renderA.c:374 contains the call to VPCompAR11B() which as you recall from the gprof profile executes frequently. In addition, the only calls to VPCompAR11B() are from this loop. Therefore, it can be inferred that the loop at vp_renderA.c:374 executes frequently as well. Naturally, any loops inside the primary *flow of control* within VPCompAR11B() would tend to execute an even larger number of times than the loop at vp_renderA.c:374. The for loop located at vp_compAR11B.c:514 iterates across the range of scanlines that make up a voxel. Finally, the while loop that starts at VPCompAR11B():line 4835 and ends at line 5097 consumes 98.94% of the execution time samples that occur inside of the function. This loop iterates over and processes the pixels that comprise a scanline and computes opacity and color of the pixel that comprises the two-dimensional intermediate image.

Table 7.10: AMIDE loop classification

#	Loop (file:line)	Type	Data
1	vp_renderA.c:374	for loop iterates by 1 or –1 across slices	Two-dimensional slices of voxels
2	vp_compAR11B.c:514	for loop iterates by 1 from 0 to variable jcount which is an argument to VPCompAR11B()	One-dimensional scanline of voxels
3	vp_compAR11B.c:4835	while loop with multiple exits, iterates based upon variable count which is decremented to zero	Voxel, Intermediate image

The three outermost loops mentioned above are counted loops with iteration counts that are bound before the loop is executed. The while loop iterates over an array of values of type GrayIntPixel which contains two floats, a short, and two unsigned chars.

On the surface, one or more of these three loops seem appropriate targets for the threading effort.

It can be inferred that the innermost loops execute many more times than the outermost loops. One concern with applying threading is the degree of overhead incurred starting up threads and assigning work. It may beneficial to thread the outermost loop and pay the expense of thread startup and shutdown as few times as possible.

7.1.3 Design and Implement

The design effort begins with the determination of the type of decomposition to apply in order to divide the computation and data. The high degree of loop-based processing on the image data suggests data decomposition is a natural fit. OpenMP is the logical choice for well-structured loops and data decomposition. As noted in the previous section, three loops contain the majority of the execution time and therefore are the candidates for where the work is divided. Figure 7.10 shows the source code for the various loops.

There are several questions to consider in determining which of the three loops should be the basis for dividing the work:

- Is the loop easily amenable to OpenMP syntax? How much source code modification is required to enact the threading?

- What's the data access pattern? Which data needs to be shared and which data needs to be private, firstprivate, and lastprivate? Are there performance concerns?

```
/* Loop #1 vp_renderA.c:374 */
for (k = kstart; k != kstop; k += kincr)

/* Loop #2 vp_compAR11B.c:514 */
for (j = 0; j <= jcount; j++) {

/* Loop #3 vp_compAR11B.c:4835 */
while (count > 0)
```

Figure 7.10: AMIDE threading candidates

- Is the thread start and stop overhead a factor if the parallel region is inside one of the loops?

Loop #1 requires modification to enable use of the OpenMP for directive because the relational operation, "!=", is not an acceptable operation as dictated by the OpenMP specification. This loop uses the "!=" relational operation because the loop may either count up from `kstart` to `kstop` or count down from `kstart` to `kstop` depending on the direction specified by the variable, `vpc->reverse_slice_order`. Loop #2 can employ an OpenMP for directive in a straightforward fashion with no source modification. Loop #3 requires modification because the loop is a while loop and contains multiple exits. OpenMP does not have a work sharing directive that can be used on a while loop. In addition, OpenMP requires a parallel region to be well structured and not have multiple exits.

Data access is a primary concern for any threading effort because there are both correctness and performance issues that can occur if the threading is not properly implemented. Transforming a three-dimensional set of voxels into a two-dimensional image requires tracing rays through the three-dimensional data compositing a final color for a pixel in a two-dimensional image. The three-dimensional voxel data is primarily only read so potential data race issues in the voxel data are averted. Compositing of the image on the two-dimensional image requires the computation and assignment of pixel data. Two potential issues should be considered when threading the algorithm:

1. Does the processing order of the slices impact the resulting image?

2. How is access to the two-dimensional image data synchronized?

The voxel data structure (*vpc*) contains the rendering context in its entirety and will be shared between the threads. This data structure contains the voxels making up the three-dimensional image data, the intermediate image used in the shear transform as well as the final two-dimensional image data. Several values will need to be private per thread; these values are typically dependent on the iteration variable. Techniques to find these data values will be discussed in Section 7.1.4. The three loops that are under consideration for threading primarily access the voxel data and then assign to the two-dimensional intermediate image.

The thread start and stop overhead is minimized in many mature OpenMP implementations through the use of a thread pool that keeps a team of threads to use as various parallel regions are encountered. A less mature implementation would create and destroy threads

```
/* Loop #1 vp_renderA.c:374 */
#pragma omp parallel for
for (k = kstart; k < kstop; k += kincr)
```
Figure 7.11: AMIDE OpenMP modifications #1

as parallel regions are entered and exited. Even so, there is overhead in acquiring threads from a thread pool. The conclusion is that this overhead may disadvantage Loop #3 and then Loop #2, respectively because of their particular nesting level.

You may think that if threading is such a good thing, why not place OpenMP directives around all three loops? This is a reasonable idea; however, one that should be approached carefully. In general, if all available processor cores exhibit good utilization, increasing the number of threads can lead to an oversubscription situation which may be detrimental to performance.

Based upon the analysis of the three loops, Loop #1 will be considered for parallelization. Implementation is fairly straightforward and Figure 7.11 shows the initial OpenMP pragmas that will be used for Loop #1.

Loop #1 was modified by changing the relational operation controlling termination of the loop from "!=" to "<". To guarantee that the loop has a positive increment, vp_view.c:256 is modified so vpc->reverse_slice_order=0. In a production implementation, it would be necessary to replicate the loop and loop contents with one for loop used for positive values of kincr and one for loop used for negative values of kincr. For the sake of this case study, I choose to ignore the issue and only concern myself with one direction.

Figure 7.12 lists the build options required to build with OpenMP. The execution of AMIDE with the modification to Loop #1 seems to execute correctly. Visually, the images appear to be similar to the original execution although I do not implement an exhaustive test to determine if the executions are truly equivalent. This stresses an important point during application development and that is the need for a robust test suite. In addition, in a graphical application whose output is dependent on floating point calculations it is unreasonable to compare the grayscale values of each pixel between the original and optimized versions. Instead, in a production environment you would compute a signal-to-noise calculation between pixels in the original and optimized version and specify in advance a value that must not be exceeded to meet appropriate quality. This value would be determined manually by inspecting the effect on image quality by different noise levels.

Configure and build VolPack	`./configure CC=icc CXX=icc CFLAGS="-openmp"` `make`

Configure and build AMIDE	`./configure CC=icc CXX=icc LD=xild LDFLAGS="-openmp`
	`-L/<build directory>/volpack-1.0c4/lib"`
	`make CFLAGS="-I/<build directory>/volpack-1.0c4/include"`

Set environment	`export LD_LIBRARY_PATH="$LD_LIBRARY_PATH:/<build` `directory>/volpack-1.0c4/lib"`

Figure 7.12: AMIDE OpenMP build options

At this point, it may be tempting to rush into performance tuning with the Loop #1 modification; however, some unresolved debug issues need to be addressed.

7.1.4 Debug

Even though the modifications to Loop #1 resulted in no apparent issues, it still makes sense to perform the classification of variables step to help reduce the chance for thread issues outside of your normal testing. This section details the steps in debugging Loop #1 by properly classifying data access and then using a tool to help find difficult to diagnose issues lurking in the code.

Classifying data access focuses on finding variables that should be placed in OpenMP private and firstprivate clauses. Recall, by default all variables are shared so as far as stability is concerned, there is no need to explicitly label shared variables. It may benefit performance-wise to privatize a shared variable.

Recall that a variable will need to be declared private if its value is loop dependent, for example, its value is dependent on the loop iteration variable. A variable will need to be declared firstprivate if the variable is initialized before the parallel region and if it is dependent on the loop iteration variable.

7.1.4.1 AMIDE Loop #1 Debug

The application with the Loop #1 modifications seemed to execute correctly; however, analysis of the enclosed parallel for region shows a number of variables whose values

```
/* Loop #1 vp_renderA.c:374 */
#pragma omp parallel for private (shadow_slice_u, shadow_slice_v, shadow_slice_u_int,
shadow_slice_v_int, shadow_slice_start_index, shadow_image, slice_u_frac,
slice_v_frac, WgtTL, WgtBL, WgtTR, WgtBR, scan_offset_index, run_lengths,
voxel_data, slice_u, slice_v, slice_u_int, slice_v_int, slice_start_index, intimage)
firstprivate (shadow_k, slice_depth_cueing)
for (k = kstart; k < kstop; k += kincr)
```

Figure 7.13: AMIDE Loop #1 modification 2

Set memory use limits	`export TC_OPTIONS="heap=2000000000"`
Execute AMIDE under Intel® Thread Checker	`tcheck_cl .libs/lt-amide ../m2862-small.xif`

Figure 7.14: Executing Intel® Thread Checker on AMIDE

are dependent on the loop iteration variable and subsequently used in the loop. These variables should be made private. At `vp_renderA.c:423` is a reference to the loop iteration variable, *k*, and a number of variables set dependent upon it. The dependent variables should be declared in a private clause on the OpenMP parallel for directive in order to prevent data race issues. For example, two threads may set the value of `slice_u` before the subsequent read uses the correct thread-specific value. An analysis of the dependent variables suggests the modification to the OpenMP pragma detailed in Figure 7.13. The variables, `shadow_k` and `slice_depth_cueing`, are declared in a firstprivate clause because of their initialization before the OpenMP parallel region.

In order to prove the benefit of the second set of modifications when there is no apparent difference in execution time behavior, Intel® Thread Checker is used to corroborate the need for the privatization. In other words, Intel® Thread Checker should find thread errors on the application using the modifications from Figure 7.11 that do not exist when using the second set of modifications. The debug build of the application is executed by Intel® Thread Checker using the steps detailed in Figure 7.14.

The first option limits the memory use of Intel® Thread Checker to approximately 2 GB. The particular system used during the debug phase had 2.5 GB of RAM available. Due to the overhead of instrumentation, each execution of AMIDE took several hours, even with the preceding modification.

One helpful tip when employing Intel® Thread Checker is to reduce the working set of the application. AMIDE was modified to limit the iterations and rotations of the image

ID	Short Des cription	Se ve ri ty Na me	C o u n t	Context [Best]	Description	1st Ac cess [B est]	2nd Acc ess [Bes t]
2	Write -> Read data-race	Er ro r	6	"vp_ren derA.c" :377	Memory read at "vp_renderA.c":423 conflicts with a prior memory write at "vp_renderA.c":423 (flow dependence)	"vp_re nderA. c":423	"vp_ren derA.c" :423
5	Read -> Write data-race	Er ro r	6	"vp_ren derA.c" :377	Memory write at "vp_renderA.c":427 conflicts with a prior memory read at "vp_renderA.c":438 (anti dependence)	"vp_re nderA. c":438	"vp_ren derA.c" :427
6	Write -> Write data-race	Er ro r	6	"vp_ren derA.c" :377	Memory write at "vp_renderA.c":428 conflicts with a prior memory write at "vp_renderA.c":428 (output dependence)	"vp_re nderA. c":428	"vp_ren derA.c" :428
8	Write -> Write data-race	Er ro r	6	"vp_ren derA.c" :377	Memory write at "vp_renderA.c":429 conflicts with a prior memory write at "vp_renderA.c":429 (output dependence)	"vp_re nderA. c":429	"vp_ren derA.c" :429

Figure 7.15: AMIDE Loop #1 Intel® Thread Checker output

to two instead of the previously used forty iterations employed by the benchmark. A subset of the output from the Intel® Thread Checker run on the application with the modifications from Figure 7.11 is detailed in Figure 7.15. Error ID 2 specifies a data race involving a read and write of variable `shadow_w` at `vp_renderA.c:423`. Error ID 5 specifies a data race involving a read of `slice_u` at `vp_renderA.c:438` and a write of the value at `vp_renderA.c:427`. Intel® Thread Checker is then executed on the application with the modification proposed in Figure 7.13. The errors associated with the variables listed in the private and firstprivate clauses no longer occur. Unfortunately, not all of the errors have been eliminated.

Figure 7.16 shows a subset of the remaining errors. Further analysis of the error listing reveals an issue around reads and writes to values pointed to by the variable *ipixel*. The variable *ipixel* references values in the two-dimensional intermediate image. Error ID 2 specifies a data race between a read occurring at `vp_compAR11B.c:5058` and a

ID	Short Description	Severity Name	Count	Context [Best]	Description	1st Access [Best]	2nd Access [Best]
2	Write -> Read data-race	Error	5481 490	"vp_compAR11B.c":2 40	Memory read at "vp_compAR11B.c":5058 conflicts with a prior memory write at "vp_compAR11B.c":5065 (flow dependence)	"vp_compAR11B.c":5065	"vp_compAR11B.c":5058
4	Write -> Write data-race	Error	5481 495	"vp_compAR11B.c":2 40	Memory write at "vp_compAR11B.c":5063 conflicts with a prior memory write at "vp_compAR11B.c":5063 (output dependence)	"vp_compAR11B.c":5063	"vp_compAR11B.c":5063
5	Write -> Write data-race	Error	5481 493	"vp_compAR11B.c":2 40	Memory write at "vp_compAR11B.c":5065 conflicts with a prior memory write at "vp_compAR11B.c":5065 (output dependence)	"vp_compAR11B.c":5065	"vp_compAR11B.c":5065
7	Write -> Read data-race	Error	217	"vp_compAR11B.c":2 40	Memory read at "vp_compAR11B.c":4844 conflicts with a prior memory write at "vp_compAR11B.c":5088 (flow dependence)	"vp_compAR11B.c":5088	"vp_compAR11B.c":4844
8	Write -> Read data-race	Error	217	"vp_compAR11B.c":2 40	Memory read at "vp_compAR11B.c":668 conflicts with a prior memory write at "vp_compAR11B.c":5088 (flow dependence)	"vp_compAR11B.c":5088	"vp_compAR11B.c":668

Figure 7.16: AMIDE Loop #1 modification 2 Intel® Thread Checker output

write that occurs at `vp_compAR11B.c:5065` on the value pointed at by `ipixel->opcflt`. The code in this region is attempting to read, modify, and write a specific pixel without synchronization so that it is possible that another thread is accessing the data value at the same time. This is a classic data race and is similar to the issue illustrated in Figure 6.4. Error ID 7 and Error ID 8 specify similar data races occurring on `ipixel->lnk`. The `lnk` value is used to perform early ray termination and is a performance optimization.

7.1.4.2 Resolving Debug Issues

There is no quick fix that will address the issue around concurrent access to dereferences of *ipixel*. There are two options to consider in resolving the data race issues:

1. Add synchronization around every dereference of ipixel.

2. Restructure the code to eliminate the need for concurrent access to dereferences of ipixel.

Option #1 is not worth considering. The execution profile shows the region of code where these dereferences occur are the most frequently executed regions of code. Adding synchronization inside the most frequently executed regions of code would significantly impact performance. This option also illustrates the difference between *coarse-grain locking* and *fine-grain locking*. One method of addressing the data race is to place a lock around access to *ipixel*. This is an example of a coarse-grain lock which would solve the issue, but result in penalizing access for all dereferences of *ipixel* because the issue only exists when the value of *ipixel* in two threads are the same. A fine-grain lock approach would associate a mutex with each and every possible pixel on the two-dimensional image. Such a solution would increase the memory use of the application and be difficult to code correctly.

Option #2 requires significant modification of the code and is summarized as follows:

- Create a private copy of the intermediate image, `vpc->int_image.gray_intim`, for every thread.

- After the parallel loop has executed, the private copies of the intermediate image are merged together to produce the final reduced version of the intermediate image.

The vpc data structure is modified by adding a new field, `tempZ`, of type (GrayIntPixel**) which will hold an array of per thread intermediate images. The function, `VPResizeRenderBuffers()`, is modified to allocate and deallocate the `tempZ` field similar to `vpc->int_image.gray_intim`. The size of the `tempZ` array is determined by calling `omp_get_num_procs()`, which determines the number of processors on the system and subsequently is used to determine the number of threads to create when executing the parallel region. Be sure to allocate and deallocate the memory properly; my first implementation only allocated the memory. The application would work correctly for most of the benchmark and then rendering of the images would stop once the system memory was exhausted.

Inside the parallel region, `int image` is set equal to

`&vpc->tempZ[omp_get_thread_num()][slice_start_index]`

which is the intermediate image that is subsequently passed to `composite_func()`. This modification sets the intermediate image to the thread-specific memory space for use in the call to `composite_func()`. Each thread is determining its own value for the intermediate image based upon the slices it is analyzing. After the parallel region, the collection of partial results needs to be reduced to one value. A loop is added after the OpenMP parallel for directive which iterates through the thread-specific intermediate images and combines the per pixel opacity and color values using the same algorithm employed in the `VPCompAR11B()`. Code to perform this reduction is listed in Figure 7.17. Due to the parallel execution and reduction, the order of computation is not the same as the serial version and since floating point operations are being computed, reordering the operations may not result in the same precise pixel values as the serial execution.[2] The transformation, however, results in images that appear visually similar and is therefore a justifiable modification.

The issue regarding concurrent access to `ipixel->lnk` is a bit easier to resolve. One option is to synchronize read and write access to every occurrence of `ipixel->lnk`, but the overhead of synchronizing access would be too great. Instead, the code contains directives which allow turning off of the functionality (removes the assignments to `lnk`)

```
for (thread_iter = 1; thread_iter < tempZ_num_threads; thread_iter++) {
  int dim_threadz = vpc->intermediate_width * vpc->intermediate_height;
  int tempz_iter;
  for (tempz_iter=0;tempz_iter<dim_threadz;tempz_iter++) {
    if (vpc->tempZ[thread_iter][tempz_iter].opcflt > vpc->min_opacity) {
      float iopc, iopc_inv;
      iopc = vpc->tempZ[0][tempz_iter].opcflt;
      iopc_inv = (float)1. - iopc;
      vpc->tempZ[0][tempz_iter].clrflt += vpc->tempZ[thread_iter][tempz_iter].clrflt *
                                          iopc_inv;
      iopc += vpc->tempZ[thread_iter][tempz_iter].opcflt * iopc_inv;
      vpc->tempZ[0][tempz_iter].opcflt = iopc;
    }
  }
}
vpc->int_image.gray_intim = vpc->tempZ[0];
```

Figure 7.17: AMIDE intermediate image reduction

---

[2] A good read on fallacies involving floating-point numbers is: http://www.zuras.net/FPFallacies.html

and therefore prevents these issues from occurring in the first place. The build settings are modified by adding the following to the VolPack compile command (CFLAGS):

`-DSKIP_ERT`

A comparison between the application executing the original benchmark with and without the build option, `-DSKIP_ERT`, was performed and showed no significant difference in execution time.

Intel® Thread Checker was then executed on the application with the modifications to address the data race issues and the errors associated with `ipixel` were no longer present.

7.1.5 Tune

The tuning phase focuses on the performance of the threaded implementation compared to the serial implementation. Timing results for the serial implementation at two optimization levels and the threaded implementation using one, two, and four threads, respectively are listed in Table 7.11. The application is executed on a system containing two Dual-Core Intel Xeon® Processors LV for a total of four processor cores.

Speedup is computed by dividing the baseline time results by the time results for a particular execution. In this case, the baseline is the serial version of AMIDE built with icc 10.0 and the "`-O3 -xP`" optimization settings.

The threaded versions were modified to employ the number of threads specified by the environment variable, `OMP_NUM_THREADS` and set equal to one, two, and four, respectively.

Table 7.11: AMIDE parallel optimization times

Result	Compiler	Optimization setting	Time result (in seconds)	Speedup
1	icc 9.1	`-O2`	100.735	0.967
2	icc 10.0	`-O3 -xP`	97.452	1.00
3	icc 9.1	`-O2 -openmp` `One thread`	107.913	0.903
4	icc 9.1	`-O2 -openmp` `Two threads`	78.309	1.244
5	icc 9.1	`-O2 -openmp` `Four threads`	55.866	1.744

The following observations can be made on the performance information:

- The use of OpenMP incurs execution time overhead of approximately 7%. This is apparent by comparing result #1 and result #3.

- Speedup of the threaded version using two threads is 1.244 as indicated by result #4.

- Speedup of the threaded version using four threads is 1.744 as indicated by result #5.

Table 7.12 is a comparison between the obtained speedup detailed in Table 7.11 and the theoretical speedup calculated in Table 7.9. Obtained speedup has certainly improved; however, the theoretical limit suggests there may be headroom to increase performance even further.

Intel® Thread Profiler is employed to profile the debug build of the application and provide clues on potential performance improvements. The steps to create the profile for the AMIDE benchmark are summarized in Table 7.13.

The command line version of Intel® Thread Profiler is employed on Linux to collect the profile which is stored in the directory ./threadprofiler. The directory is archived and copied for viewing by the Windows version of Intel® Thread Profiler.

Profile generation using Intel® Thread Profiler has the issue as experienced on VTune™ Analyzer in that the generated profile is of the entire application, which includes startup

Table 7.12: AMIDE speedup-obtained versus theoretical

Number of threads	Obtained speedup	Theoretical speedup
Two threads	1.244	1.89
Four threads	1.744	3.39

Table 7.13: Intel® Thread Profiler steps

Steps	Commands
Source environment	`source /opt/intel/itt/tprofile/bin/32/ tprofilevars.sh`
Execute command line version on AMIDE	`tprofile_cl .libs/lt-amide ../ m2862-small.xif`
Archive the profile directory and copy to system executing Windows	`tar -cvf tprofile.tar ./threadprofiler`

and user input. Intel® Thread Profiler allows you to zoom into the area of interest which in this case is the benchmark. Figure 7.18 is a screenshot of the profile results zoomed into the time period of the volume rendering. The profile indicates that all four processors were utilized 70.54% of the time during this period.

A performance issue can be observed by zooming into the Timeline view of the profile. Figure 7.19 shows the timeline from time period 100 to 110, and one observation is the blocks of spin time for threads three, four, and five occurring after each fully utilized execution. This characteristic is indicative of a potential workload balance issue that can be solved through the use of a scheduling clause on the OpenMP work sharing directive.

A number of different scheduling types and block sizes were executed on the application to see which one provided the greatest performance improvement. Figure 7.20 shows the modification to the AMIDE Loop #1 to use a schedule clause with a dynamic distribution and a block size of 10.

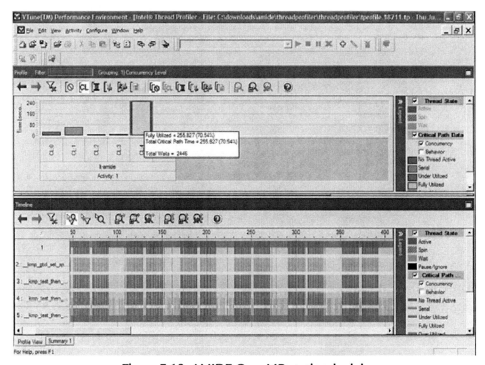

Figure 7.18: AMIDE OpenMP static schedule

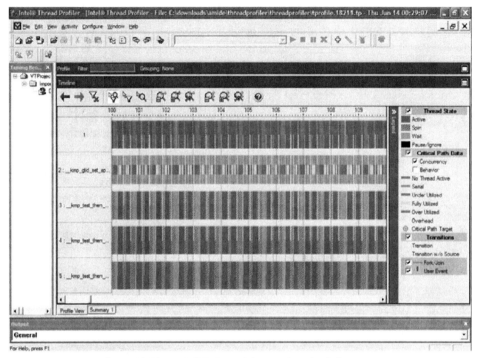

Figure 7.19: AMIDE OpenMP static schedule zoom

```
/* Loop #1 vp_renderA.c:374 */
#pragma omp parallel for private (shadow_slice_u, shadow_slice_v, shadow_slice_u_int,
shadow_slice_v_int, shadow_slice_start_index, shadow_image, slice_u_frac,
slice_v_frac, WgtTL, WgtBL, WgtTR, WgtBR, scan_offset_index, run_lengths,
voxel_data, slice_u, slice_v, slice_u_int, slice_v_int, slice_start_index, intimage)
firstprivate (shadow_k, slice_depth_cueing) schedule (dynamic, 10)
for (k = kstart; k < kstop; k += kincr)
```

Figure 7.20: AMIDE Loop #1 modification #3

Table 7.14 shows the timing results of using two different scheduling clauses. A modest improvement of approximately 5% results from employing both dynamic and static scheduling with a block size of 10.

At this point, one additional change is measured and that is the use of advanced optimizations including high general optimization (-O3), automatic vectorization (-xP), and interprocedural analysis (-ip) which further improves the execution time reducing

Table 7.14: AMIDE performance tuning

Compiler	Optimization setting number of threads (if >1)	Time result (in seconds)	Speedup
icc 10.0	`-O3 -xP`	97.452	1.00
icc 9.1	`-O2 -openmp, four threads`	55.866	1.744
icc 9.1	`-O2 -openmp, four threads schedule (static, 10)`	52.805	1.845
icc 9.1	`-O2 -openmp, four threads schedule (dynamic, 10)`	52.268	1.864
icc 10.0	`-O3 -xP -ip -openmp, four threads, schedule (dynamic, 10)`	42.984	2.267

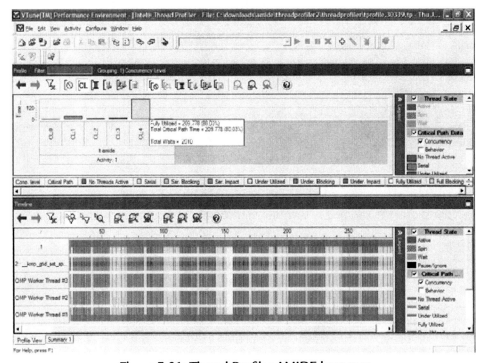

Figure 7.21: Thread Profiler AMIDE best case

it to 42.984 s and providing a speedup over the baseline serial optimization of 2.267. We have now achieved a reasonable speedup employing multi-core processors. Figure 7.21 is a screenshot of a profile using the scheduling clause with advanced optimization and shows the system is fully utilized 80.03% of the execution time of the benchmark.

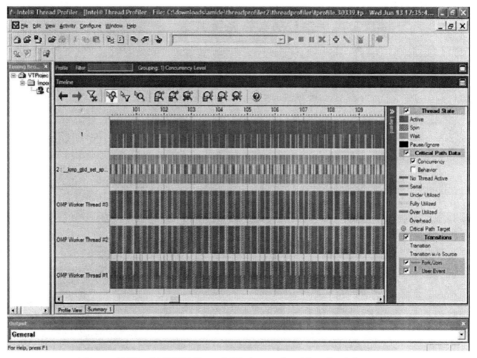

Figure 7.22: AMIDE OpenMP dynamic, 10 schedule zoom

Figure 7.22 shows the zoomed in timeline view with less wait time compared to the previous timeline view.

At this point, the case study is complete. A speedup of 2.267 has been achieved by threading.

A good inclusion to close a case study is to suggest future areas for investigation if further improvement is desired. One interesting benchmark behavior is observed in Table 7.15. The time for each individual rotation ranges between 1.9 and 3.5 s for the Y-axis rotations and 1.5 and 1.9 s for the X-axis rotations. The Y-axis rotations execute slower than the X-axis rotations.

The EBS profile has already provided valuable information on the application; however, EBS is capable of profiling on many more events and can help suggest common

Table 7.15: *Y*-axis versus *X*-axis timing results

Rotation	Y-axis	X-axis
1	1.916	1.6
2	1.943	1.578
3	1.957	1.595
4	3.546	1.926
5	3.582	1.944
6	3.452	1.934
7	3.49	1.948
8	3.599	1.935
9	1.95	1.59
10	1.927	1.58
11	1.91	1.598
12	1.903	1.606
13	3.429	1.952
14	3.447	1.925
15	3.538	1.9
16	3.444	1.905
17	3.47	1.942
18	1.901	1.609
19	1.921	1.618
20	1.921	1.581

performance issues. Since the application is iterating over large amounts of voxel data, memory utilization and cache effectiveness is a natural area to investigate.

VTune™ Analyzer is employed to collect profile information for two memory-related events, Level 1 Data Cache misses and Level 2 Cache misses. These events are collected using the Event Ratio feature of VTune™ Analyzer which collects profiles of multiple events and creates a ratio of the numbers. Figure 7.23 is a screenshot of selecting the L2 Miss per Instruction ratio which is comprised of two events, Level 2 Cache misses and Instructions Retired. A profile is also generated on L1 D-Cache Miss per Instruction Retired which is comprised of L1 D-Cache misses and Instructions Retired.

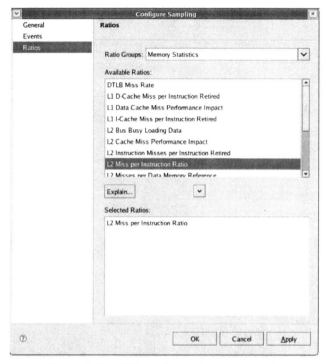

Figure 7.23: EBS configuration–event ratio

Three versions of the benchmark were executed and EBS profiles collected on each:

1. Original benchmark
2. Benchmark modified to perform *X*-axis Rotations Only
3. Benchmark modified to perform *Y*-axis Rotations Only.

The modified benchmarks performed the same absolute number of rotations as the original benchmark; however, only performed one type of rotation, *X*-axis or *Y*-axis.

The VTune™ Analyzer instruction manual specifies the ratio ranges as good and bad as shown in Table 7.16.

Table 7.17 summarizes the EBS ratio results obtained by profiling the three runs. Notice the *Y*-axis rotations have a higher L1 misses per instruction retired and L2 misses per

Table 7.16: Cache ratio categorization

	Good	Bad
L1 ratio	<0.020	>0.080
L2 ratio	<0.001	>0.003

Table 7.17: Cache EBS profile

	L1 misses per instruction retired	L2 misses per instruction retired
Benchmark	0.0167	0.0018
X-axis Rotations Only	0.0126	0.0001
Y-axis Rotations Only	0.0207	0.0034

instruction retired. The measurements for the L2 Misses per Instruction Retired for the *Y*-axis rotations are also in the bad category. Future analysis should discover why the miss rate is high and what can be done to lower it.

This case study was effective in detailing the required actions at each step of the TDC, however, similar to real development projects, there is never enough time to complete every possible step. Instead these items are suggested areas for future analysis and optimization and are summarized as:

- Improving the data set by including additional images to gauge performance. This step also includes improving the input data for profile-guided optimization.

- Study the effects of adding an OpenMP parallel for construct to the loop denoted Loop #2.

- Study scaling effects by increasing the number of threads or the number of processor cores.

Chapter Summary

This chapter provides a case study of multi-threading AMIDE by employing the TDC. First, a benchmark program of the application was developed that was optimized for serial execution. Analysis employed performance analysis tools to determine the most frequently executed portions of the application. Design and development employed

OpenMP as the threading API. The debugging phase revealed several data race issues, which were resolved. Finally, the tuning phase optimized thread balance and produced an execution time speedup of 2.267 on a four processor core system.

References

[1]   AMIDE, http://amide.sourceforege.net
[2]   VolPack Volume Rendering Library, http://graphics.stanford.edu/software/volpack/
[3]   AMIDE sample, mouse_2862, http://amide.sourceforge.net/samples.html
[4]   P. Lacroute and M. Levoy, Fast Volume Rendering Using a Shear-Warp Factorization of the Viewing Transformation, *Computer Graphics Proceedings*, SIGGRAPH, 1994.
[5]   AMIDE download, http://sourceforge.net/projects/amide
[6]   Volpack download, http://sourceforge.net/projects/amide/

## CHAPTER 8

Case Study: Functional Decomposition

Arun Raghunath

Key Points

- Multi-threaded applications perform poorly if the method of decomposition results in excessive synchronization overhead or inefficient cache behavior. The performance results for the multi-threaded version of the application may even be worse than the serial version of the application.

- A call graph profile provides insight into how to effectively functionally decompose an application.

- Some thread errors may be errors in the theoretical sense, but not in the practical sense. Application knowledge is required to make the correct assessment before implementing unnecessary synchronization.

The previous chapter applied the threading development cycle (TDC) to an application whose concurrent work was distributed based upon data decomposition. There are other classes of applications that effectively decompose based upon the data; however, some applications benefit from a different decomposition technique. This chapter employs the TDC to parallelize an application where the concurrent work is found via functional decomposition. Like the previous chapter, each step in the process is documented with details on the thought process, tools usage, and results. I intend to focus more upon the technique used to decompose the application as the first technique applied resulted in poor performance and a redesign of the threading is required to obtain performance benefit. The expectation is that you, the reader, are provided enough detail to understand the motivation and application so that the techniques can be applied to your threading project. For this case study, the speedup on a set of input traces ranged from 1.02 to 6.27, the performance of the original single-threaded application executing on a system with eight processor cores.

8.1 Snort

Snort[*] is a well-known open-source intrusion detection system which performs deep packet inspection of network traffic and alerts system owners of suspicious behavior based on user configurable rules. Snort is a user-space application. Both versions of the application used in this case study, versions 2.2.0 and 2.4.4, are single threaded.

8.1.1 Application Overview

This section provides a high level overview of the basic operation of Snort. Figure 8.1 illustrates the high-level processing stages of Snort and describes the main modules showing how a packet is handled as it flows through the application. Snort dataflow begins by a packet being read off of the network. Snort uses the packet capture library [1] (libpcap[1]) to read in packets from user space. Next, the packet is decoded to determine its type. This stage identifies the link level protocol (e.g., Ethernet) as well as network (IP) and transport layer protocols (TCP, UDP). The protocol headers are verified and packets with malformed headers are handled appropriately. Following the packet capture stage is the preprocessor stage, which consists of a set of filters that identify potentially dangerous packets and pass this information on to the detection engine. The detection engine looks for specific patterns in the packets. Analysis behavior of the detection engine is entirely configurable by the user, who must specify "rules" that the engine applies to the packets. Each rule specifies a pattern that the detection engine should look for and the action to perform when a pattern is detected in the packet. General actions include passing a packet through, logging information about the packet, and dropping the packet. In addition, an alert can be raised at the console or sent over a network socket to a system administrator on another machine.

---

[1] libpcap is an open-source library that provides a system-independent interface for user-level packet capture.

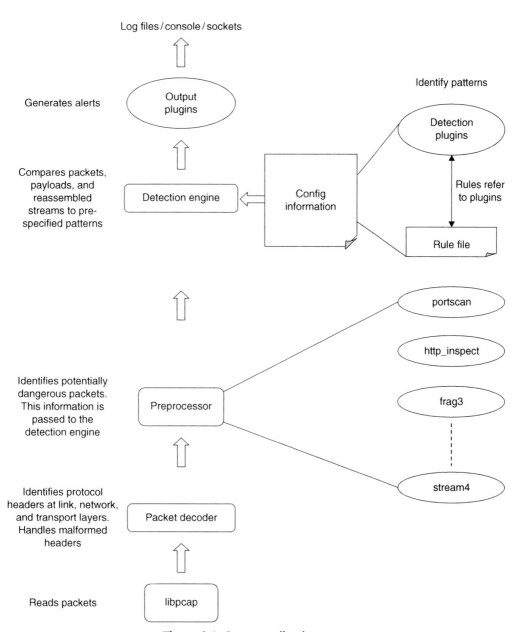

Figure 8.1: Snort application steps

Snort provides a well-defined interface for each of the stages of the packet processing described above. Programmers can create their own plugins that adhere to this interface and insert them into a specific stage in the packet dataflow. This ability allows the intrusion detection capabilities of Snort to be easily extended to handle new kinds of attacks. It also makes the code modular as the code specific to an attack is isolated in a specific plugin. Snort comes with numerous predefined plugins some of which are described below to provide the reader a better feel for how Snort operates:

- Portscan plugin – contains code to detect portscan attacks. It maintains statistics of connection attempts such as the source address, destination addresses, the destination ports, etc. over a specified time interval.

- Http_inspect plugin – monitors packets containing web requests.

- Frag3 plugin – processes fragmented IP packets.

- Stream4 plugin – reassembles packets comprising a TCP session. The reassembled stream is then fed in to the detection engine. This allows the engine to not only look for potential attack patterns deep inside a packet but across packets as well.

Attack detection rules refer to snippets of code termed detection plugins. This allows programmers to specify their own code to identify a desired pattern. Like the preprocessors, the detection plugins can be inserted into the packet dataflow in Snort simply by tweaking the Snort configuration. Finally, the action to be taken when a pattern is matched is also configurable via output plugins, thereby allowing the user to control how Snort behaves when a specific attack is detected.

8.1.2 Build Procedure

The basic build procedure for Snort is documented in Table 8.1. The steps consist of downloading Snort, setting up the build directories, and compiling it.

Removing the current build of Snort in preparation for a new build is accomplished by typing:

```
make clean
```

in the snort-2.4.4 directory. The optimization process employs different values for CC and CFLAGS which enable the use of different compilers and optimization

Table 8.1: Snort build procedure

Download and untar files[a]	`mkdir <build directory>`
	`download snort-2.4.4.tar.gz`
	`tar -zxvf snort-2.4.4.tar.gz`
	(libpcap is a prerequisite for installation.)
Configure and build Snort	`cd snort-2.4.4`
	`./configure`
	`make CC=gcc CFLAGS+=-O2`
	`make install`
Execute Snort with input trace file	`$HOME/snort-2.4.4/src/snort -r <trace file>`
	`-N -l <log directory>`

[a] Snort versions 2.2.0 and 2.4.4 were used for the case study. These instructions should still work for updates; however, some modifications may be required.

settings; these differences will be detailed at each step. Execution of Snort requires specifying a trace file as the input file[2] and a log directory which specifies a directory for the output.

8.2 Analysis

The analysis of Snort is preparation for the eventual parallel optimization effort and includes serial optimization and a study of the application to determine the best locations to thread.

8.2.1 Serial Optimization

Before threading the application, two compilers and several compiler optimizations are employed on the serial version of the code. This serial optimization is a prerequisite for parallel optimization and serves a dual purpose. First, the serial optimization provides a baseline of performance for the parallel optimization effort and second, the optimization effort provides familiarity with the application. The serial optimization steps and results are summarized in Table 8.2.

---

[2] This is true when reading packets from a file, not if packets are coming from the network.

Table 8.2: Snort serial optimization summary

Serial optimization step	Results
Characterize the application	Streaming data access, integer operations
Prioritize compiler optimization	Memory optimizations, interprocedural optimizations
Select benchmark	Processing of packet traces using default rules
Evaluate performance of compiler optimizations	Best results were obtained using `-ipo -prof-use` with a gain of 11.23% over baseline.

Characterization of the application reveals the following:

- Snort exhibits a *streaming access* pattern over large amounts of packet data. This suggests that data access optimizations may prove beneficial.

- Snort is predominately an integer application which suggests automatic vectorization may not help a great deal.

- Optimizations that benefit cache utilization, including profile-guided optimization that typically benefits code size, may improve performance of the application.

The Snort application is very large, consisting of numerous modules to perform operations such as packet decoding, stream reassembly, rule matching, attack detection, and alert generation. There are numerous data structures including flow state variables and statistical counters that are shared across the modules. The code to perform these operations is integer in nature, not floating-point.

The compiler optimizations to apply should target data placement optimization. Therefore `-O3` is prioritized high. Since the application is processing large amounts of packet data, data and instruction cache optimizations may provide a performance boost. Profile-guided optimization will be used with the expectation of benefiting application code size. Using the compiler optimization settings, an 11.23% execution time improvement was obtained over the baseline optimization settings. Code size of the best-performing version was 1.57% smaller than the baseline, which highlights the code size reduction abilities of profile-guided optimization. The next few sections provide further detail on each phase of serial optimization.

8.2.2 Benchmark

A sample of timing output from an execution of the benchmark appears in Figure 8.2. Timing output is represented by the line:

```
Snort total run time was 0.162837 seconds
```

The experiments in this case study used either Snort version 2.2 and Snort version 2.4.4 and the default intrusion detection rule set included with the specific version. In order to benchmark Snort performance, real world packet traces are used for input consisting of headers obtained from NLANR [2] and Intel's Information Technology department with synthetically generated payloads to test the updated version of Snort. The HTTP payloads were generated based on the recorded surfing of a user to many popular websites, while the other protocol payloads were created using random data. The trace files were stored and read as a file from disk instead of employing actual input from the network; this approach reduces I/O latency and increases reproducibility. Code to record time at the start and end of packet processing was added and the difference between them is then output. Little variance was observed when executing the application compiled at a given optimization setting.

8.2.3 Serial Optimization Results

The serial optimization phase employs multiple compiler optimization levels to improve the performance of the application. The compiler, optimization settings, and time results are summarized in Table 8.3. The relative performance column is the percentage difference for the average of the time to process two traces using gcc at -O2 as the baseline. The time for the processing of a trace is the average of 10 executions[3] of Snort using the trace as input. Positive values for relative performance equate to better performance. The relative code size column is the ratio of code and data segment size between the application compiled by a given compiler and optimization settings and the application compiled by gcc at -O2 as the baseline. The system where the tests were executed is an Intel® Core™ 2 Duo processor with a clock speed of 2.66 GHz, 2.5 GB RAM, and executing the Red Hat Enterprise Linux 4 OS.

The optimization settings were added as options to the *make* command used when building Snort. Two different compilers were employed, gcc 3.4.6 and the Intel C++ Compiler

---

[3] At a minimum, 12 runs were actually measured with the highest and lowest time discarded.

```
========================================================
Snort processed 143973 packets.
========================================================
Breakdown by protocol:
   TCP: 125650    (87.273%)
   UDP: 16963     (11.782%)
   ICMP: 240      (0.167%)
   ARP: 0         (0.000%)
   EAPOL: 0       (0.000%)
   IPv6: 0        (0.000%)
 ETHLOOP: 0       (0.000%)
   IPX: 0         (0.000%)
   FRAG: 1077     (0.748%)
   OTHER: 43      (0.030%)
 DISCARD: 0       (0.000%)
========================================================
Action Stats:
ALERTS: 0

LOGGED: 143707
PASSED: 0
========================================================
Fragmentation Stats:
Fragmented IP Packets: 1077    (0.748%)
  Fragment Trackers: 0
  Rebuilt IP Packets: 0
  Frag elements used: 0
Discarded(incomplete): 0
  Discarded(timeout): 0
  Frag2 memory faults: 0
========================================================
Snort exiting
Snort total run time was 0.162837 seconds
```

Figure 8.2: Sample Snort benchmark output

versions 10.0. GNU gcc 3.4.6 is the default compiler on the operating system. Table 8.4 summarizes the commands to build using icc 10.0 and the options -ipo and -prof-use.

One observation from the performance results is the sensitivity to different optimization settings. For example, the application compiled using -O3 and the application compiled using profile-guided optimization (-prof-use) exhibit a performance improvement

Table 8.3: Snort serial optimization settings and results

Compiler	Optimization setting	Time result (in seconds)	Relative performance improvement (%)	Relative code size increase (%)
gcc 3.4.6 (default)	-O2	0.173	0	0
icc 10.0	-O2	0.175	-2.31	8.49
icc 10.0	-O3	0.170	3.45	8.49
icc 10.0	-xO	0.184	-6.34	8.93
icc 10.0	-prof-use	0.169	1.99	2.14
icc 10.0	-ipo	0.166	4.12	4.80
icc 10.0	-O3 -prof-use	0.236	-26.73	2.17
icc 10.0	-ipo -prof-use	0.155	11.23	-1.57
icc 10.0	-ipo -O3 -prof-use	0.179	3.46	-1.52

Table 8.4: Snort ICC command line

Configure and build Snort with -prof-gen	`./configure` `make CC=icc CFLAGS+=-prof-gen CFLAGS+=-O2`
Execute Snort with input trace	`snort -r $HOME/snort/traces/ANL-1107413100-1.pcap.0.0.out -N -1 win32/`

Rebuild Snort with -prof-use	`make clean`
	`./configure`
	`make CC=icc CFLAGS+=-prof-use CFLAGS+=-ipo`

over the baseline; however, the application compiled using both -O3 and profile-guided optimization exhibits a relatively large performance decrease. For the purposes of the case study, this behavior is a curiosity that will be ignored; however, a comprehensive performance optimization would include analysis of this behavior by performing:

- Characterization of the performance by applying EBS and comparing items such as code size, data cache miss rates, branch mispredictions between several versions of the application

- Comparison of the assembly language of the hotspots between several versions of the application

8.2.4 Execution Time Profile

An execution time profile was obtained using VTune™ Performance Analyzer. Table 8.5 summarizes the system level EBS profile for clockticks obtained when executing the benchmark. As you can see, three components that comprise the Snort application consume the largest portion of time. These components are library routines in libc, kernel routines called indirectly by the application, and Snort. Further analysis[4] of the libc routines reveals `memcopy()` and `memset()` as the functions taking up the most time there.

Further analysis of the Snort component which contained 24.11% of the obtained samples reveals the functions listed in Table 8.6 as the primary time consuming functions.

From the profiles, you can conclude that there is no predominant hotspot that should be focused upon for threading; over 50% of the execution time is spent outside of the

Table 8.5: System-wide EBS profile of Snort

Name	Percentage of the process "snort"
`libc-2.3.4.so`	35.11
`vmlinux-2.6.9-42-EL`	25.74
`snort`	24.11
`scsi_mod`	9.24
`pcap_offline_read`	2.36
All other functions	3.44

Table 8.6: Process-level EBS profile of Snort

Function name	Clocktick %
`DecodeTCP`	61.90
`DecodeIP`	10.78
`DecodeRawPkt`	5.76
`CallLogPlugins`	3.01
`PcapProcessPacket`	2.76
`DecodeUDP`	2.51

---

[4] Double-clicking on individual components in the VTune Performance Analyzer GUI provides a breakdown of the percentage time spent in the individual functions comprising the larger component.

application code in library functions and the kernel. At this point, another view of program behavior is required; a call graph profile may provide insights into a target for multi-threading.

8.2.5 Call Graph Profile

Figure 8.3 is a portion of the call graph provided by an execution of VTune™ Performance Analyzer on the benchmark. `InterfaceThread()` and all of the functions that are called by it or as a result of it comprise the majority (99%) of the execution time and so the graph under `InterfaceThread()` is where focus should be spent. The functions, `pcap_loop()` and `pcap_offline_read()` reside in libpcap and are provided with a callback function into the Snort application. In this case, the function is `PcapProcessPacket()` and is responsible for further processing of the packets with child functions inside of `PcapProcessPacket()` calling the hot functions inside of Snort such as `DecodeTCP()`, `DecodeIP()`, and `DecodeRawPkt()`. A call graph does have limitations; it does not provide information on the kernel routines that from the EBS profile consume over 25% of the execution time.

Table 8.7 contains the total time and self-time for the functions depicted in Figure 8.3. Total time is the amount of time in microseconds that was spent in the particular function and all of the child functions called as a result. Self-time represents the amount of time spent in the particular function; time spent in functions called from the particular function is not included. As you can see, the two critical functions are `_IO_fread_internal()` and `PcapProcessPacket()`. The function `_IO_fread_internal()` is associated with a read to the file containing the packet data and is the one function where the majority of time is spent. The self-time of `PcapProcessPacket()` is zero; however, it contains the functions lower in the call chain that consume the remainder of the execution time. The conclusions are threefold:

- It does not make sense to attempt to multi-thread starting from only `PcapProcessPacket()` and the functions that are called as a result of it. The potential speedup obtained from threading these regions will not be attractive.

Figure 8.3: Call graph profile of Snort

Table 8.7: Call time data for Snort

Function name	Total time (microseconds)	Self-time (microseconds)
InterfaceThread	2,548,405	4
Pcap_loop	2,545,747	4
Pcap_offline_read	2,545,743	0
_IO_fread_internal	2,263,365	2,263,365
PcapProcessPacket	282,552	0
Packet_time_update	0	0
Sfthreshold_reset	0	0
ProcessPacket	282,552	0

- The target of threading will need to be higher in the call chain, higher than obviously _IO_fread_internal() as these functions are in a system library.

- A large portion of time is spent in a file I/O function that would not exist in the production version as the production version would be obtaining packets off of the network and not from a file. We should be okay for this exercise, but this stresses the need to measure the benefits on a production system at a later date.

Given the above, the most attractive threading location is InterfaceThread().

The next steps of the TDC specify creation of a flow chart of the hotspots and the classification of the most frequently executed loops. These steps are primarily relevant in cases where the hot spots are localized to one or two hotspots which then dictate where the threading is to occur. Since this characteristic is not present in Snort, there is no need to perform these steps.

8.3 Design and Implement

8.3.1 Threading Snort

Now that the serial version of Snort has been optimized and some initial characteristics of the application have been gathered, other avenues to speed up the application further are considered. One obvious route is to leverage the presence of multiple cores in the system. To do this, a multi-threaded version of Snort must be created. As long as there are parts of the Snort processing that can occur in parallel, speedup should be obtainable by having

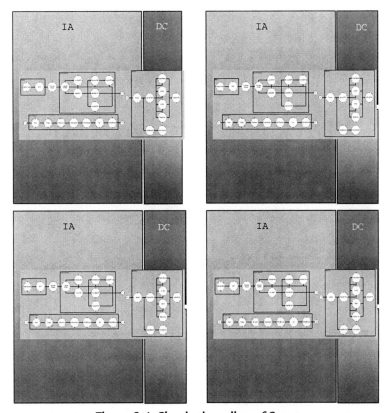

Figure 8.4: Simple threading of Snort

multiple threads work on the processing simultaneously. Since the Snort modules work with packets, it appears that multiple threads, each processing a different packet, could execute in parallel to speed up the overall processing time. In this section this approach is examined in further detail and then a performance measurement of multi-threaded Snort is detailed.

The POSIX threading library (libpthread) is employed to create the threaded version of Snort. The first approach runs the entire Snort code (minus the initialization and shutdown routines) in each thread and is referred to as simple threading. In this scenario, the number of threads created is set equal to the number of cores in the system. Each thread is set to have affinity to a specific core to prevent unnecessary thread migration, as this causes an increase in the scheduling overhead of threads. Figure 8.4 diagrams the

```
1   void *InterfaceThread(void *arg)
2   {
3       pthread_t threads[NUM_THREADS];
4       int i = 0;
5       for (i = 0; i < NUM_THREADS; ++i) {
6           if (pthread_create(&(threads[i]), NULL, SimpleThread,
7           (void*)i)){
8               ErrorMessage("pthread_create failed!\n");
9           }
10      }
11      for (i = 0; i < NUM_THREADS; ++i) {
12          if (pthread_join(threads[i], NULL)) {
13              ErrorMessage("pthread_join failed!\n");
14          }
15      }
16      CleanExit(0);
17      return NULL;
18  }
```

Figure 8.5: Code modifications 1

simple threaded version of Snort and represents four cores (IA represents the processor core and DC represents the data cache) executing four threads and each containing all of the Snort processing steps.

8.3.2 Code Modifications

Consider the code changes needed to thread Snort in the manner described above. Figure 8.5 presents the high level modifications. NUM_THREADS is a compile time variable that can be used to control the number of threads created. The function, InterfaceThread(), is modified to create an array of thread identifiers (line 3). The call to pthread_create() is used to create the desired number of threads each of which executes the function SimpleThread(). An additional argument is passed to the start function of the thread that indicates the processor core for which this thread has affinity. The call to pthread_create() returns a thread identifier, which can be used by the creating parent thread to synchronize execution with the child threads it creates (lines 5–10). Once the call to `pthread_create()` completes, a new thread of execution begins executing the start function (`SimpleThread()`. The main thread running the `InterfaceThread()` function now waits for the threads it spawned to complete by invoking the `pthread_join()` function (lines 11–15). This is a blocking call which returns when the specified thread is destroyed. Once all the threads have completed, the usual Snort shutdown routine, `CleanExit()`, is invoked before function exit.

The `SimpleThread()` function documented in Figure 8.6 is invoked with an argument that indicates the processor core on which the thread running this function should execute. The function uses this argument to create the appropriate mask for a system call[5] to set the affinity of this thread (lines 7–12). Afterward, the function initializes the thread local state and then enters a while loop which it exits only when the application is about to stop.

In each iteration of the loop the function handles one packet. It uses the libpcap to read in packets (lines 38–42). The pcap library is not thread-safe and since numerous threads execute this same function, the calls into libpcap need to be protected by locks. The mutex abstraction provided by the Pthreads library is used to create the lock (line 4) and acquire/release (lines 20–24, 55–59) it around the critical section where the pcap calls are invoked.

While the locks are needed for correct operation, they do limit the parallelism as all threads except the one that has acquired the lock are unable to perform any useful work while waiting for the lock. To increase parallelism, rather than holding the lock for the entire duration of the packet processing, a copy of the packet is created (lines 52–54) and the lock is immediately released so that other threads might proceed (lines 55–59). The copied packet information is passed to the `ProcessPacket()` function of Snort from where all the packet processing is performed. Each thread works with a different packet so the overall Snort processing can advance in parallel.

The function exits the loop (lines 25–34) when there are no more packets or if the user requests the application to be terminated. This is indicated by the setting of the *stop_threads* flag. During execution each thread maintains its own copy of the statistics for the packets it processes. On completion, these statistics are consolidated

---

[5] The modifications to Snort described here were done for a Linux system. `sched_set_affinity()` is a Linux system call that indicates the affinity of the calling thread to the scheduler.

```
1    static int pkt_count = 0;
2    static int stop_threads = 0;
3    /* We need a lock for PCAP calls since PCAP is not thread safe */
4    static pthread_mutex_t pcap_lock = PTHREAD_MUTEX_INITIALIZER;

5    void* SimpleThread(void* arg)
6    {
7      int cpu_num = (int)arg;
8      int cpu_mask = 1 << cpu_num;
9      if (sched_setaffinity(0, sizeof(cpu_mask), &cpu_mask)){
10       printf("Could not set affinity.\n");
11       pthread_exit(NULL);
12     }
13     SnortEventqInit();
14     OtnXMatchDataInitialize();
15     while (1){
16       const u_char* pkt_data;
17       u_char pkt_data_copy[4096];
18       struct pcap_pkthdr* pkt_header;
19       struct pcap_pkthdr pkt_header_copy;
20       /* Lock the count and stop variables and the pcap reader */
21       if (pthread_mutex_lock(&pcap_lock)){
22         ErrorMessage("pthread_mutex_lock failed!\n");
23         pthread_exit(NULL);
24       }
25       /* Check and see if the count is maxed out */
26       if (stop_threads ||
27       (pv.pkt_cnt != -1 && pkt_count >= pv.pkt_cnt)){
28           /* Unlock the count/reader */
29         if (pthread_mutex_unlock(&pcap_lock)){
30           ErrorMessage("pthread_mutex_unlock failed!\n");
31           pthread_exit(NULL);
32         }
33         break;
34       }
35       /* Get a new packet */
36       int err = 0;
37       pkt_data = NULL;
38       while (!err){
39         /* pcap_next_ex returns 0 if packets are being read from a
40            Live capture and a timeout occurs. */
41         err = pcap_next_ex(pd, &pkt_header, &pkt_data);
```

Figure 8.6: Function SimpleThread() (*Continued*)

```
42      }
43      if (!pkt_data){
44        stop_threads = 1;
45        /* Unlock the count/reader */
46        if (pthread_mutex_unlock(&pcap_lock)){
47          ErrorMessage("pthread_mutex_unlock failed!\n");
48          pthread_exit(NULL);
49        }
50        break;
51      }
52      memcpy(pkt_data_copy, pkt_data, pkt_header->caplen);
53      memcpy(&pkt_header_copy, pkt_header,
54            sizeof(pkt_header_copy));
55      /* Unlock the count/reader/stop */
56      if (pthread_mutex_unlock(&pcap_lock)){
57        ErrorMessage("pthread_mutex_unlock failed!\n");
58        pthread_exit(NULL);
59      }
60      /* Now run the packet through the rest of the code */
61      ProcessPacket(NULL, &pkt_header_copy, pkt_data_copy);
62    }
63    /* Combine the packet counts from the thread-local counters */
64    if (pthread_mutex_lock(&pc_mutex)){
65      ErrorMessage("pthread_mutex_lock failed!\n");
66      pthread_exit(NULL);
67    }
68    collect_stats(&global_pc, &pc);
69    if (pthread_mutex_unlock(&pc_mutex)){
70      ErrorMessage("pthread_mutex_unlock failed!\n");
71      pthread_exit(NULL);
72    }
73    return NULL;
74  }
```

Figure 8.6: (*Continued*)

into a set of global statistics variables. Since every thread needs to access these variables on exiting the loop, the function acquires another lock that protects the global statistics-related variables. The lock is released after this thread's values are added to the global state (lines 63–72).

The modifications detailed so far would be all that are needed to thread Snort if the Snort processing were only on a per packet basis; however, this is not the case. Snort

Table 8.8: Sharing categories of variables in Snort

Category	Description
Communication	The state is shared between entities that have a *producer/consumer* relationship.
Global	The state is shared between entities across packets, flows, and modules. There can be only one copy of this state across the entire application.
Thread local	The state is shared between entities across packets, flows, and modules. However, the state can be replicated amongst multiple threads with no two threads trying to access the same replica.
Flow local	The state is shared between different modules processing packets of the same flow.

maintains state for tracking sessions, connection attempts, reassembling TCP streams in addition to all of the state needed for the attack detection rule database. This state is read and written by various modules spread all over the Snort code base. Thus, the creation of a threaded version of Snort requires careful understanding of which state variables are truly independent and which are shared between different parts of the Snort code and could therefore be accessed from different threads concurrently.

A major part of the effort to thread Snort is to understand the maintained state and its sharing pattern. Table 8.8 provides an overview of the broad categories in which the shared state can be classified and how to handle each category.

Communication variables are used to pass state information between modules for a particular packet. Since each thread handles all the processing of a packet, these variables can be declared thread local and avoid the need for locking altogether. Each thread will have its own copy for communicating the state between modules.

Global state must be protected by locks as access to these from one thread or module must be protected. High synchronization costs could be incurred if multiple threads attempt to access these at the same time.

The thread local category comprises global state that can be replicated. An example is the set of variables that maintain statistics such as the count of TCP segments, IP fragments, etc. These counts can be maintained separately per thread and then reduced to one global value later. So, all such variables are converted to be local to a thread. When the thread is exiting, the local statistics generated by that thread are summed into the global statistics (access to this will have to be protected by locks as explained earlier).

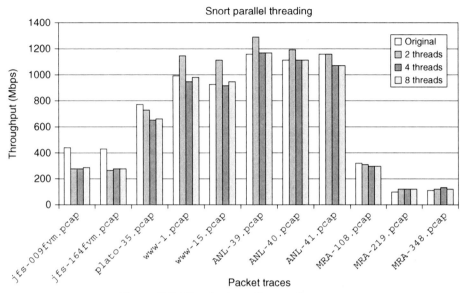

Figure 8.7: Simple threading performance

The flow local category comprises variables that are all being accessed and updated while processing packets of a *flow*. In the current application structure where any thread can be processing any packet regardless of its flow, access to these variables must be protected with a lock.

All of the shared (i.e., global) variables of the original single-threaded Snort application were then categorized as described above, analyzing the semantics of the variables and the sharing patterns and accordingly choosing to either lock access to the variables or replicate them across threads. Replicating the variables per thread has the benefit of reducing synchronization overheads and potentially improving cache locality.

Figure 8.7 is a performance comparison between the original version of Snort 2.2[6] and the simple threaded versions at a number of different threads. Throughput is calculated by dividing the size of the trace in bits by the execution time and is reported as Megabits per second. These results were obtained from executing the application on a dual processor

---

[6] Snort 2.2 was employed due to timing issues. The simple threading work was performed early in the case study timeline; at that time version 2.2 was the most up-to-date version.

Table 8.9: Simple threading performance results

Version	Throughput (Megabits per second)	L2 Cache hit rate (%)
Original	566	99
Four thread	543	86

system where each processor has four cores. For a wide variety of input traces, there was no appreciable performance gain; in fact, in many of the cases the performance dropped as a result of employing the multi-threaded version executing on multi-core processors.

For example, on the `jfs-164fvm.pcap` trace, the original, single-threaded version of Snort exhibited throughput of approximately 400 Megabits per second (Mbps). The 2 thread, 4 thread, and 8 thread versions of Snort executing on the 8 processor core system exhibited performance of approximately 250 Mbps.

The obvious question from these results is: why does the performance drop on several of the traces as a result of the multi-threading? One suspicion is cache performance; perhaps the data associated with individual packet flows is migrating from core to core as a result of context switching. An EBS profile provides insight into the answer. Consider the performance results detailed in Table 8.9 that are correlated with the L2 cache hit rate returned from an EBS profile of the single-threaded and simple threaded version. L2 cache hit rate was computed by taking the ratio of EBS-obtained counts on two architectural events, L2 Cache Request Misses and L2 Cache Requests and subtracting the ratio from one. The EBS profile was obtained using traces which contained on the order of 175 concurrent flows and executed on a system with four processor cores. In a production environment you would want to obtain performance numbers on the planned target; however, sometimes analysis (and accessibility) can be easier on a system with fewer cores.

As you can see, the throughput between the original and the simple-threaded (four threads) versions are 566 Mbps and 543 Mbps, while the L2 Cache Hit Rate is 99% versus 86%. It is difficult to reach any conclusions based upon the cache hit rate; however, the decrease in cache hit rate does suggest some amount of *cache thrashing* is occurring.

Alternatively, obtaining a *thread profile* of the application provides insight into the poor performance results. The first step required is a reduction of the input data. Initially, a sample input using a trace file containing over 1,000,000 packets was used; however, due to the memory-intensive nature of profiling, the memory limits were reached on

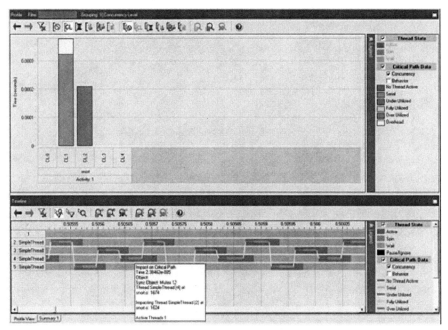

Figure 8.8: Thread profile of simple threaded Snort

the system. A trace converter was applied to limit the trace file to 100,000 packets and the command line version of Intel® Thread Profiler was employed to create the profile. Similar to the thread profile of AMIDE, the command line version of the tool was used to obtain the profile on a system using a Linux OS. The profile was then transferred to a Windows-based system for analysis. A screenshot of the obtained profile is depicted in Figure 8.8. The profile reports average concurrency of 1.33 when executing on a system with four processor cores; this result is very disappointing. The timeline view of the profile reveals large amounts of serial time caused by contention on a mutex at lines 1674 and 1624 of `snort.c`. This mutex serializes access to the PCAP library and routines because PCAP is not thread-safe.

The analysis of the multi-threaded version of the Snort code shows that the potential performance gains from multiple cores are not apparent due to the overheads of locking and cache thrashing. So is all the effort wasted? Certainly not! In the next section other characteristics of the application are leveraged to extract parallelism in a different manner that allows Snort performance to improve.

8.3.3 Flow Pinning

In this section the concept of flow pinning is discussed which includes:

- The definition of a flow,

- How a flow impacts the operation of Snort

- How flows are used to improve the performance of Snort

As stated earlier, Snort operates on distinct packets and as such parallelization of Snort can be accomplished by processing distinct packets at the same time. However, quite often the processing of many of the distinct packets involves access to the same state (i.e., the same memory location). An obvious example of this is a TCP session which is a logical connection between two endpoints that stream data between each other. This "stream of data" is sent over the network as a set of distinct packets. At the sender side, the data stream is broken into smaller chunks that can fit into the maximum size packet[7] that can be transmitted between the endpoints. At the receiver side, these distinct packets are picked up, the packet headers are stripped, and the original data stream is reassembled by placing the packet payload at the correct offset in a separate buffer. It is this recreated buffer that is sent to the end application. In order to place the data at the right offset, the receiving side needs to maintain state (a contiguous buffer in memory, offsets, and other counters and flags to identify the connection status). In order to process any packet containing data for a particular TCP session, the receiving side needs to either read or write this state. All such distinct packets that need access to some common state for their processing are said to be part of the same flow.

The problem with access to flow state when parallelizing the application is that the different threads (running on different cores) attempt to access the same flow state concurrently. For correct operation, access to the state must be synchronized using locks. This however impacts the ability to extract parallelism as now the parallel threads of execution end up wasting time trying to acquire a lock that is currently held by another thread. In effect these other threads cannot do any useful packet processing work during this time. This is the lock synchronization overhead that was discussed in the previous attempt to parallelize Snort by using simple threading.

---

[7] Networks impose a physical limit on the maximum size of data that can be transported on it as one unit. When an application needs to send more data than this physical maximum, packetization needs to occur.

The second problem with multiple simultaneous threads trying to access the same flow state is the impact on the caches. When a thread running on a core acquires the lock on flow state and begins to read and write the state, this state gets pulled into the processor core's cache. Subsequently another thread executing on a different processor core performs the same operations and pulls the same state into its cache. When this happens, the state is evicted from the first processor core's cache. Consequently, the next time the first processor core tries to access the state it incurs a cache miss and must wait for the state to again be pulled back into its cache. This is an instance of cache thrashing and occurs to every core that attempts to access the flow state. As was stated in the earlier analysis, this is the second major reason for the lack of scaling of the performance of the simple threaded version of Snort.

Both the problems associated with accessing flow state have to do with multiple threads of execution having to access the same flow state simultaneously. The consequence of this is that not only is the processing serialized (only the one thread which acquired the lock can proceed), but the single-threaded performance is reduced as well due to cache thrashing. To avoid these issues, a property of the workload seen by the Snort application is exploited. Typically the devices running an intrusion detection system such as Snort are placed at the boundary of an administrative domain (like the front end of a datacenter or at the edge of an enterprise network). Consequently the packets that are seen by the device at any given point in time belong to numerous flows, all of which are simultaneously active. Moreover, the state of each flow is independent of the other, which implies that the states of two different flows can be concurrently accessed and modified. This *flow level parallelism* can therefore be exploited to improve the performance of Snort.

To leverage flow level parallelism, only one thread should process all of the packets associated with a flow. Doing this eliminates the locking overheads (as there is only one contender for the state) as well as the cache thrashing (the state that is cached in one core can remain there unless it is evicted due to other data accessed by the same core). Moreover, the expectation is that the workload consists of multiple simultaneous flows and can ensure that each thread has a separate flow to process, thereby allowing all of the threads (and their corresponding processor cores) to be doing useful work. In other words, the packets of different flows must be separate and all of the packets of a given flow should be processed by the same processor core. This technique is called *flow pinning* as the flow is in effect "pinned" to one processor core.

Flow Pinning

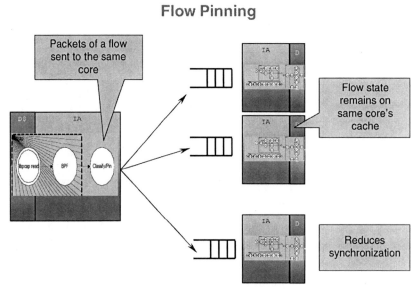

Figure 8.9: Snort flow pinning

Flow pinning is enabled by inserting a new "classifier" stage to the Snort packet dataflow. The classifier stage determines which flow a packet belongs to and accordingly determines which processor core should process the packet. In addition, a fundamental change is made to the threading model. One thread is created that runs the pcap and the classifier stage while the remaining threads execute the rest of the code. As before, each thread has affinity set to one processor core. A queue is created for each of these remaining threads. The first thread (pcap+classifier) reads packets in, determines which processor core should process it, and places the packet into the queue for the thread running on that processor core. The remaining threads run in a loop, picking packets from their respective queues, completing the required processing for a packet, and returning back to pick up the next packet from the queue. Figure 8.9 illustrates the structure of the modified application.

Functionally, decomposing the application into separate threads provides some benefits. Recall that the pcap library is single threaded and hence the simple threaded version requires serialization of the calls to this library using a lock. As a result, when one thread acquires the lock the other threads trying to acquire the lock are stalled. This overhead is avoided by dedicating a thread for calling into the pcap library to obtain packets.

Second, the classifier was introduced to ensure packets of a flow are processed on the same processor core. If the earlier model was followed, which allowed every thread to classify packets and place the packets into the appropriate core's queue, there would be contention for access to the queues. When two (or more) threads classify packets to the same core concurrently, they would need to serialize access to the queue. This would again mean the threads would have to stall waiting for their turn to get exclusive access to the queue. Further, the lock variable and the queue head pointer would be constantly moving between the caches of the processor cores running the thread trying to add a packet to the queue. This is because each thread needs exclusive access to these variables, which requires the previous cached copies to be invalidated each time a new thread gains access to the queue.

By decomposing the Snort dataflow functionally as explained earlier all of these issues are avoided:

- There is no need for locks around the access to the pcap library as there is only one thread making calls to the pcap library.

- There is no contention for access to the queues between the classifier and the rest of the code as there is always one producer (the pcap+classify stage) and one consumer (the one thread responsible for running the rest of the stages on a given core).

8.3.4 Code Modifications for Flow Pinning

The code needed to enable the application structure that was just described will now be detailed. The first noteworthy point is that due to the change in the threading model, there is only one thread reading in the packet data. The input packets are then passed to the other threads for processing. This implies that packets allocated by one thread need to be handed off to the other threads. The straightforward approach of allocating memory from the heap using system calls would be too expensive (kernel-user mode transitions per packet). Instead, the implementation takes advantage of the fact that Snort reads in a fixed amount (4K) of the packet data to pre-allocate a pool of packet buffers. A queue with pointers to these buffers is then populated. The cost of allocating a packet is now reduced to acquiring a buffer pointer from the queue. When Snort is done processing a packet, the packet pointer is simply returned to the queue. Synchronization costs between

```
1    void *InterfaceThread(void *arg)
2    {
3      pthread_t process_stage_threads[NUM_PROCESS_THREADS];
4      pthread_t pcap_stage_thread;
5      if (packet_pool_init()){
6        ErrorMessage("Could not initialize packet pool!\n");
7      }
8      for (int i = 0; i < NUM_PROCESS_THREADS; ++i){
9        pcap_to_process_queues[i]=
10         ring_queue_create(NO_SYNCH,NO_SYNCH,NO_SYNCH,NULL);
11       if (!pcap_to_process_queues[i]){
12         ErrorMessage("Could not create processing queue!\n");
13       }
14     }
15     if (pthread_spin_init(&pc_spinlock, 0)){
16       ErrorMessage("InterfaceThread: pthread_spin_init failed!\n");
17     }
18     for (i = 0; i < NUM_PROCESS_THREADS; ++i){
19       if (pthread_create(&(process_stage_threads[i]),
20            NULL,ProcessStageThread,(void*)i)){
21         ErrorMessage("pthread_create failed!\n");
22       }
23     }
24     if (pthread_create(&pcap_stage_thread,
25          NULL,PcapStageThread,(void*)0)){
26       ErrorMessage("pthread_create failed!\n");
27     }
28     for (i = 0; i < NUM_PROCESS_THREADS; ++i){
29       if (pthread_join(process_stage_threads[i], NULL)){
30         ErrorMessage("pthread_join failed!\n");
31       }
32     }
33     if (pthread_join(pcap_stage_thread, NULL)){
34       ErrorMessage("pthread_join failed!\n");
35     }
36     if (packet_pool_shutdown()){
37       ErrorMessage("Could not shutdown packet pool!\n");
38     }
```

Figure 8.10: Function `InterfaceThread()` (*Continued*)

```
39    for (i = 0; i < NUM_PROCESS_THREADS; ++i){
40      queue_destroy(pcap_to_process_queues[i]);
41    }
42    CleanExit(0);
43    return NULL;
44  }
```

Figure 8.10: (*Continued*)

all of the processing threads trying to access one global packet pool is avoided by the use of a per thread packet pool.

In the code included in the following figures, the stage containing the code to read in packets from the pcap library and classify the packets is termed the Pcap stage. The stage containing the rest of Snort processing is referred to as the Process stage. A compile time variable, *NUM_PROCESS_THREADS*, defines the number of threads that will run the Process stage, while *NUM_CPUS* defines the number of available processor cores in the system.

Three new functions are defined: `InterfaceThread()`, `PcapStageThread()`, and `ProcessStageThread()`. Consider the function `InterfaceThread()` detailed in Figure 8.10.

This function begins by initializing the packet pool in lines 5–7. Lines 8–14 create the queues that are needed between the Pcap stage and the Process stage. One queue is created per thread that runs the Process stage. Consider the function `ring_queue_create()` whose parameters are detailed in Figure 8.11 and which is responsible for creating the queue.

The function `ring_queue_create()` creates a ring buffer of a fixed size and returns a pointer to the queue created for use in subsequent queue calls. In order to optimize access to the queue, this queue implementation allows the programmer to specify the kind of synchronization to use for the operations supported by the queue. The synchronization options supported are mutex, spinlocks, and no synchronization. Depending on the usage, the programmer can choose the options that result in the most efficient queue implementation being chosen. For example, if the operation is expected to complete quickly, a spinlock is very efficient. On the other hand if the operation is expected to take a long time (like when a queue is usually full and threads are trying to write to it) then a

```
queue_t* ring_queue_create(synch_type_t read_synch,
                           synch_type_t write_synch,
                           synch_type_t get_len_synch,
                           drop_fn drop);
```

read_synch	The type of synchronization to use while reading from the queue
write_synch	The type of synchronization to use while writing from the queue
get_len_synch	The type of synchronization to use while getting the queue length
drop	The function to use when determining which element in the queue should be dropped when the queue overflows. If set to NULL, the default behavior is to block when a write is attempted when the queue is full

Figure 8.11: Parameters for `ring_queue_create()`

mutex is likely more efficient as it allows the scheduler to choose other threads to run in place of the thread that is waiting to write.

In the flow pinning implementation, no synchronization is needed while reading, writing, or obtaining the length of the queue because the application structure ensures that there is always only one reader and one writer for the queues. Hence there is no point in creating and acquiring locks for the read and write operations. The function to obtain the queue length is never called, so no locks or mutexes are created for it. Finally, the default drop behavior was chosen, which will result in writes to a full queue to block until there is space in the queue.

Lines 15–17 create a global spinlock that is used by the threads to synchronize access to the global statistics collected by Snort. Lines 18–23 create the threads to run the Process stage. The start function for these threads is `ProcessStageThread()` which will be discussed in more detail shortly. Lines 24–27 create the one and only thread that executes the Pcap stage by passing in the `PcapStageThread()` as the start function. On return from the call to `pthread_create()`, a new thread is created that executes the start function. The other parameter passed into the call to `pthread_create()` is the argument with which the start function is invoked. In both the call to `pthread_call()`

to spawn `ProcessStageThread()` and the call to `pthread_call()` to spawn
`PCapStageThread()`, a number is passed that the spawned functions then use to
determine the processor core on which it should execute. Thus, after executing the code
up to line 36 the main thread will have spawned numerous threads running the Pcap and
the Process stages on the appropriate processor cores.

At this point, the `InterfaceThread()` function waits for the other threads it
spawned to complete by invoking the `pthread_join()` call on each of the spawned
threads (lines 28–35). The `pthread_join()` function blocks until the thread on
which it is called completes, at which point it returns the value returned by the completing
thread to the caller. Once all of the threads have completed, `InterfaceThread()`
shuts down the packet pool, destroys the queues it created earlier, and calls the Snort exit
function, `CleanExit()`, which cleans up all the application state, and depending on
the arguments with which Snort was invoked, displays all the statistics gathered by Snort
during the execution. If there is any error while performing any of the operations described
above, the `InterfaceThread()` function displays an error message and exits.

The `PcapStageThread()` function listed in Figure 8.12 is invoked with an
argument which indicates the processor core on which the thread running this function
should execute. The function uses this argument to create the appropriate mask for a
system call[8] to set the affinity of this thread (lines 5–10). Next, the function enters a
while loop which it exits only when the application is about to stop. In each iteration of
the loop, the function handles one packet. Line 14 shows the libpcap call to the function
`pcap_next()` to read the next available packet. If there is no more data or if the
application was otherwise instructed by the user to stop (the *stop_threads* flag is set
in this case) the function exits the loop and the thread is destroyed.

When the libpcap returns a packet, the function first determines which core should
process the packet by calling the function, `get_process_queue_num()` (line 20),
which is explained in detail later. A new packet buffer is obtained from the packet pool
for the chosen core and the packet header and the payload are copied into it (lines 21–27).
A pointer to this cloned packet is then written to the chosen core's queue. As explained
in the `InterfaceThread()` function earlier, an array of queues is created for this

---

[8] The modifications to Snort described here were done for a Linux system. `sched_set_affinity` is a
Linux system call that indicates the affinity of the calling thread to the scheduler.

```
1   static queue_t* pcap_to_process_queues[NUM_PROCESS_THREADS];
2   void* PcapStageThread(void* arg)
3   {
4     struct pcap_pkthdr PktHdr;
5     int threadId = (int)arg;
6     unsigned long cpu_mask = 1 << threadId;
7     if (sched_setaffinity(0, sizeof(cpu_mask), &cpu_mask)){
8       printf("Could not set affinity.\n");
9       pthread_exit(NULL);
10    }
11    while (1){
12      const u_char* pkt_data;
13      Packet* pkt = NULL;
14      pkt_data = pcap_next(pd, &PktHdr);
15      if (!pkt_data || stop_threads){
16        pkt = NULL;
17        stop_threads = 1;
18        break;
19      }
20      int queue_index = get_process_queue_num(pkt_data, datalink);
21      pkt = allocate_packet(queue_index);
22      if (!pkt){
23        ErrorMessage("Could not allocate packet!\n");
24        pthread_exit(NULL);
25      }
26      memcpy(pkt->pkth, &PktHdr, sizeof(struct pcap_pkthdr));
27      memcpy(pkt->pkt, pkt_data, pkt->pkth->caplen);
28      queue_t* queue = pcap_to_process_queues[queue_index];
29      if (queue->write(queue, pkt)){
30        ErrorMessage("Could not enqueue packet for processing!\n");
31        pthread_exit(NULL);
32      }
33    }
34    return NULL;
35  }
```

Figure 8.12: Function **PcapStageThread()**

```
1    #define PRIME_NUMBER 251
2    int get_process_queue_num(const char* pkt, int datalink_type)
3    {
4      static int lookup_table_inited = 0;
5      static int lookup_table[PRIME_NUMBER];
6      if (!lookup_table_inited){
7        lookup_table_inited = 1;
8        for (int i = 0; i < PRIME_NUMBER; ++i){
9          lookup_table[i] = (int)((long long int)rand() *
10            NUM_PROCESS_THREADS / RAND_MAX);
11       }
12     }
13     IPHdr* ip = (IPHdr*)((char*)pkt);
14     if (ip->ip_proto == IPPROTO_TCP){
15       int hlen = IP_HLEN(ip) * 4;
16       TCPHdr* tcp = (TCPHdr*)((char*)ip + hlen);
17       unsigned int hash = ((unsigned int)tcp->th_sport) ^
18                           ((unsigned int)tcp->th_dport) ^
19                           ((unsigned int)ip->ip_src.s_addr) ^
20                           ((unsigned int)ip->ip_dst.s_addr);
21       hash = hash % PRIME_NUMBER;
22       return lookup_table[hash];
23     } else if (ip->ip_proto == IPPROTO_UDP){
24       unsigned int hash = ((unsigned int)ip->ip_src.s_addr) ^
25                           ((unsigned int)ip->ip_dst.s_addr);
26       hash = hash % PRIME_NUMBER;
27       return lookup_table[hash];
28     }
29     /*If the packet is not TCP or UDP, process it on thread 0 */
30     return 0;
31   }
```

Figure 8.13: Function `get_process_queue_num()`

communication between the Pcap and the Process stages. The packet needs to be copied because the libpcap is not thread-safe and reuses memory. The current packet must be processed by another thread while this thread continues executing the Pcap stage by looping back to read the next packet using libpcap.

The next step is to decide the flow which a packet belongs to and the core which must process it. The get_process_queue_num() function listed in Figure 8.13

implements a very simple classifier to illustrate the concept. It maintains a lookup table initialized with random numbers from 0 to the number of process threads (lines 6–12). The table consists of a prime number (251) of entries.

The function `get_process_queue_num()` generates a hash value by performing an XOR operation on some fields of the packet header (lines 17–20, 24–25). The fields chosen depend on the type of the packet. For TCP packets, the four tuple of source address, destination address, source port, and destination port is used. For UDP packets, only the source and destination ports are used. This hash value is the flow identifier for the packet as all packets belonging to the same flow will end up with the same hash value.

Since there are a relatively small number of cores, the flow identifiers generated by the function must be folded into a smaller number, namely the number of process threads; a lookup table is used for this purpose. A modulo operation on the hash value is first performed to reduce it to nothing larger than the chosen prime number (lines 21 and 26). The resulting value is then used as an index into the lookup table to pick out a number between 0 and the number of process threads. The function, `get_process_queue_num()` (lines 22 & 27), returns this value.

Just like the Pcap stage, the `ProcessStageThread()` function begins by setting the affinity of the thread running it to the core indicated by its argument (lines 3–9) (Figure 8.14). It also uses the core number to determine the queue from which it should read. Lines 11–14 involve calling various routines to initialize the state for this thread. Recall that several global variables were changed to thread local variables in order to make Snort thread-safe. Consequently this thread local state needs to be appropriately initialized when a new thread is created.

After completing all initializations, the `ProcessStageThread()` function enters the main loop where it picks out (on line 16) the pointer to the packets added to this thread's queue by the Pcap stage. Each packet is then processed by calling `ProcessPacket()` (line 24). This is a function present in the original Snort code and is an entry point to the rest of the Snort processing code. Once the processing is complete, the function frees and returns the packet buffer back to the packet pool of the correct core. This loop continues as long as there are packets present in the queue and the *stop_threads* flag is not set.

After exiting the main loop, this function consolidates all the statistics gathered by the processing performed on this thread into the global statistics counters (line 31). Access

```
1   void* ProcessStageThread(void* arg)
2   {
3     int threadId = (int)arg;
4     unsigned long cpu_mask =
5       1 << ((NUM_PROCESS_THREADS*NUM_CPUS-1-threadId) % NUM_CPUS);
6     if (sched_setaffinity(0, sizeof(cpu_mask), &cpu_mask)){
7       ErrorMessage("Could not set affinity.\n");
8       pthread_exit(NULL);
9     }
10    queue_t* queue = pcap_to_process_queues[threadId];
11    SnortEventqInit();
12    OtnXMatchDataInitialize();
13    Stream4PerThreadInit();
14    FlowPerThreadInit();
15    while (1){
16      Packet* pkt = queue->read(queue);
17      if (!pkt){
18        if (stop_threads){
19          break;
20        } else {
21          continue;
22        }
23      }
24      ProcessPacket(pkt);
25      free_packet(pkt, threadId);
26    }
27    if (pthread_spin_lock(&pc_spinlock)){
28      ErrorMessage("ProcessStageThread: spin lock failed!\n");
29      pthread_exit(NULL);
30    }
31    collect_stats(&global_pc, &pc)
32    if (pthread_spin_unlock(&pc_spinlock)) {
33      ErrorMessage("ProcessStageThread: spin unlock failed!\n");
34      pthread_exit(NULL);
35    }
36    return NULL;
37  }
```

Figure 8.14: Function `ProcessStageThread()`

to the global statistics is synchronized using the global lock `pc_spinlock` (lines 27–30 and 32–35). After transferring the statistics from the thread local variables to the corresponding globals, the `ProcessStageThread()` function exits. This causes the thread to be destroyed.

In summary, the Snort application was flow pinned by:

1. Functionally decomposing the code into two logical stages

2. Adding a queue to facilitate communication between the logical stages

3. Introducing a classifier in the first stage to identify flows

4. Creating the correct number (based on the system core count) of threads each with affinity to one core and running one of the logical stages

8.4 Snort Debug

With a significant change to the program to implement multi-threading comes the need to verifying program correctness. Manual verification of 100,000 lines of C code spanning many modules with shared state and linked in with system libraries would not only be extremely time consuming but also very difficult. The application was instrumented and executed by Intel® Thread Checker, which produced a list of diagnostics containing identified threading errors. This section limits itself to sharing techniques for effective use of the thread verification tool because the various solutions for fixing the thread issues are general in nature. More important are tips for effective use of the thread verification tool which can be applied to your particular threading project.

8.4.1 Filtering Out False Errors

Examination of the diagnostics revealed several errors that reference snort.c at line 1725. An analysis of the source code revealed a problem in a call to memcpy and the memory associated with packet data. Recall that Snort was divided into the Pcap stage that classifies and then provides the particular packet to the Process stage. Once the first stage places a packet descriptor on the queue, it gives up ownership of the packet and will not attempt to access it henceforth. Once a thread running the second stage picks up a descriptor from the queue, it assumes ownership of the packet. This transfer of ownership is implicit, occurs at a logical level, and by design will not result in a threading error even though there is no explicit synchronization in the code. Effective use of the thread

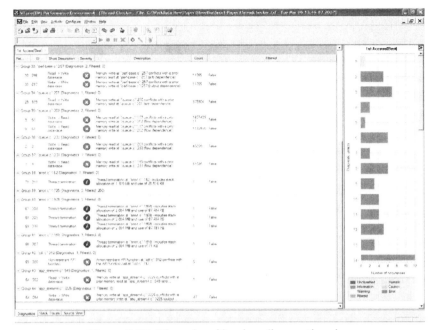

Figure 8.15: Thread verification diagnostics view

verification tool requires filtering out these false errors so that they do not appear in your diagnostic view and in the case of Intel® Thread Checker is accomplished by right clicking on the diagnostic and selecting the "Filter Diagnostic" menu option. Figure 8.15 is a screenshot of the diagnostics view using Intel® Thread Checker on the Snort application.

A second helpful view in analyzing thread errors is the source code view which allows you to view the errors correlated with source code. With the Intel® Thread Checker tool, a source view is available by double clicking a particular diagnostic. Figure 8.16 is a screenshot of a source view that highlights a data race issue on a bitmask data structure. The diagnostic is flagged as an error because the original single-threaded code used a global structure containing a bitmask to speed up the detection process. The bitmask is cleared at the start of rule matching for each session. Any time a rule matches, the appropriate bit is set to indicate that the rule has been triggered. The search algorithm uses this bitmask to avoid reevaluating rules for subsequent packets in this session.

Figure 8.16: Thread verification source view

With multiple threads executing the detection modules concurrently, this would lead to a data race. While one thread attempts to read the bitmask another thread could be resetting it. For correct operation, the bitmask needs to have valid values within a given thread. Since the rule set is static, correct operation can be achieved by replicating the structure holding the bitmask for every thread. Privatization of the data element eliminates the race condition.

Intel® Thread Checker identified several data race issues related to statistical counters, which were fixed by making the variables thread local. The other data races were fixed by guarding variables by using critical sections.

8.5 Tune

The tuning phase for Snort focuses on an assessment of the performance results to understand where flow pinning performs well. Although further performance improvement is a goal of the tuning stage, the redesign from a simple threading approach to the flow pinning approach can be considered a tuning step that required going back to the design step of the TDC. Therefore, the tune step here will focus on simply the assessment and analysis of the performance numbers. The first question to answer is the

Table 8.10: EBS profile comparison between threaded implementations

Version	175 Concurrent packet flows	25,000 Concurrent packet flows		
	Throughput (Mbps)	L2 Cache hit rate (%)	Throughput (Mbps)	L2 Cache hit rate (%)
Original	566	99	30	41
Simple threaded (four threads)	543	86	29	42
Flow pinned (four threads)	576	94	188	70

impact of keeping packets associated with a particular flow assigned to a specific core. The conjecture is that cache hit rates would be higher as a result.

An EBS profile was obtained for the flow pinned version of Snort and a comparison between them, the original version and the simple threaded version, appear in Table 8.10. The L2 cache hit rate was computed by taking the ratio of EBS-obtained counts on two architectural events, L2 Cache Request Misses and L2 Cache Requests and subtracting the ratio from one. Speedup offered by the flow pinned version is computed by dividing the throughput of the flow pinned version by the throughput of the original version. When using the input file with 175 concurrent flows, speedup of the flow pinned version is 1.02. When using the input file with 25,000 concurrent flows, speedup of the flow pinned version is 6.27.

In order to confirm the benefits of the flow pinned version compared to the simple threaded version, Figure 8.17 shows the performance across a number of packet traces. The performance improvement[9] for the flow pinned version with 8 threads over the simple threaded version with 2 threads ranged from 37% to 800%.

Figure 8.18 shows the performance results between the simple threaded and flow pinned versions of Snort on two different input traces; the result depicted on the left graph is for a trace with a large number of flows, approximately 25,000 while the result depicted on the right graph is for a trace with a small number of flows, approximately 175.

---

[9] Performance numbers obtained on a system with eight processor cores; recall though that the simple threaded version did not show scaling going from 2 to 4 and 8 cores.

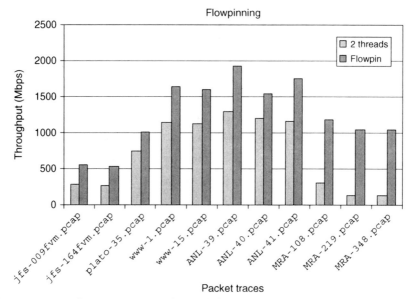

Figure 8.17: Performance comparison of simple threaded and flow pinned Snort

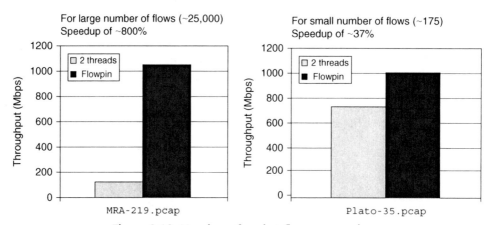

Figure 8.18: Number of packet flows comparison

A number of conclusions are reached as a result of the data:

- Flow pinning results in improved cache utilization compared to the simple threaded version (94% versus 86% and 70% versus 42%). The improvement in L2 cache hit rate is attributed to the marshaling of packets associated with a flow

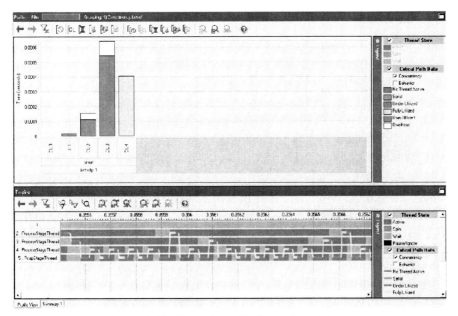

Figure 8.19: Thread profile flow pinned Snort

to the same core and the reduction of cache thrashing that would result. Better cache utilization would naturally lend itself to improved performance.

• Throughput is highest in the flow pinned version. In traces containing a larger number of concurrent packet flows (25,000), the performance improvement of the flow pinned version is substantial (188 Mbps versus 30 Mbps and 29 Mbps).

Finally, a thread profile confirms the other contributor to performance improvement, which is the removal of locks around the pcap routines and the removal of locks that result from the distributed processing of individual flows on different processor cores. Average concurrency from the profile was 3.14 on a system with four processor cores. Figure 8.19 shows a timeline view of the flow pinned implementation and a comparison with the timeline view from Figure 8.8 shows thread activity consistent with a greater degree of parallel execution.

Chapter Summary

The Snort application, an intrusion detection system, was multi-threaded to improve performance. First, the application was analyzed to determine the logical location for the threaded implementation. The hotspots of the application were spread between application, library, and kernel functions; a call graph of application execution was instrumental in finding the correct location for threading. Second, parallelization based upon data decomposition was employed to divide the processing of the incoming packets. This initial simple threaded implementation resulted in negligible performance improvement over the baseline serial version. A different approach was proposed and implemented based upon observations of the application. Flow pinning is motivated by the conjecture that better cache utilization can result by keeping data associated with related packets local to a core and by restructuring the application to allow removal of synchronization around a non-thread-safe library.

The performance results of the flow pinned version of Snort were compelling as long as the number of concurrent flows was large. On traces with a smaller number of packet flows, it is believed that additional scaling can be found. Further analysis determined that the lack of scaling is not due to workload imbalance. Instead, it is suspected that some packet flows result in code paths through the application that backup the processing of packets in the queues. Future work would include adding further threading to parallelize the hotspots that are part of these queue backups.

References

[1]   Libpcap project, http://sourceforge.net/projects/libpcap
[2]   NLANR, *Measurement & Network Analysis*, http://pma.nlanr.net/Traces/

## CHAPTER 9

Virtualization and Partitioning

Jamel Tayeb

Key Points

- Virtualization provides several capabilities including workload isolation, workload consolidation, and workload migration.

- Virtualization and partitioning combined with multi-core processors enable consolidation of embedded applications while still meeting the performance demands of each application.

- Virtualization of IA-32 architecture processors must overcome challenges involving privilege management, memory management, interrupt management, and exception management.

- VirtualLogix™ VLX for the Network Infrastructure is an example product that takes advantage of virtualization and multi-core processors.

9.1 Overview

The very existence of this chapter provides evidence of the industry's continually growing interest in virtualization. In the next 5 years, virtualization, along with its twin concept, partitioning, will become mainstream technologies in many computing environments, including the telecom and embedded market segments.

The term, virtualization, is overloaded and encompasses a large variety of concepts as is frequently the case with emerging technologies. At one end of the spectrum, virtualization can describe technologies such as byte code machines (e.g., the Java Virtual Machine). At the other end of the spectrum, virtualization can refer to the use of virtual functions in everyday developers' application coding. However, virtualization, as covered in this chapter, is a new comer in the embedded and the telecom arena. It is important to mention that the type of virtualization covered in this chapter has existed for decades in the mainframes where it was introduced in the 1960s to multiplex the use of expensive hardware resources.

In the first section of this chapter, hardware virtualization and its twin technology, hardware partitioning, are defined. In the second section, the essential use models and benefits of virtualization for embedded and telecom systems are reviewed. The third section details processor virtualization, implementation, and design considerations; this section also shows how Intel® Virtualization Technology for IA-32 Intel® Architecture (Intel® VT-x) eases their implementation. Finally, in the fourth section, a usage model of microprocessor virtualization and partitioning for the telecom market is presented by demonstrating VirtualLogix™ VLX for the Network Infrastructure solution, focusing on computing and network Input/Output (I/O) performance based on Intel® VT-x technology.

9.1.1 Preamble

Virtualization encompasses a large variety of concepts; however, a single common attribute of all these concepts is *abstraction*. By saying this, it is easier to realize why the umbrella noun, virtualization, can legitimately apply to a vast pallet of techniques. In this preamble, a discussion of virtualization is constrained to the computing environment. Let's quickly review some of the most common virtualization, or abstraction techniques in use today. Operating systems (OSes) are an example of an abstraction. One of many possible definitions of an OS is: a software layer that provides hardware abstraction and exposes unified interfaces to the applications. Indeed, many programmers are not concerned about hardware registers to be programmed or the precise timing required to perform a low-level I/O read operation depending on the type of hardware installed in their system. Instead, they desire read functionality provided by a higher level language or library application programming interface (API) and expect it to behave as specified by the standard. The same applies to the underlying libraries, making a call to an OS low-level read function. Even the OS has its own hardware abstraction layer (HAL).

To software engineers, abstraction is a well-known technique. For example, the Pascal programming language introduced the notion of pseudo-code, or P-code, in the early 1970s. P-code relies upon a byte code generation and execution technique. The compiler generates pseudo-code targeting a software stack machine with its instruction set and an execution engine – a run-time virtual machine (VM) – translates the P-code into machine code for the target platform during execution. The advantage of this approach is the disconnection of the compiler's front end and code generator from the target architecture, making it trivial to port the compiler to a new processor architecture. In this scenario, only the run-time environment has to be re-written. Today, this is exactly the same technology foundation that is used in the Java and .NET frameworks. A standard language and environment are used to develop applications, which are interpreted or compiled into an intermediate language. This intermediate code is then executed by a VM (Java Virtual Machine, Common Language Runtime, or Parrot). The .NET environment extends the abstraction by allowing many high-level languages to be translated into Microsoft Intermediate Language (MSIL), and then compiled into Common Intermediate Language (CIL) byte code. As complex and sophisticated as these technologies sound, they are based on a concept introduced circa 1960.

Consider the case where a binary targets a specific instruction set architecture (ISA) and OS, and needs to execute on a different ISA and OS combination. Typically porting the source code requires employing the correct compiler to recompile and generate a new binary. However, this is not the only way; you can also use the services of a binary translator. Such software decodes the instructions in the binary and translates them dynamically into the instructions for the targeted ISA and OS. Many binary translation software products exist; IA32EL, Aries Binary Translator, QEMU, QuickTransit, and Wabi are just a few examples.

A binary translator is capable of dynamically translating binary code for an ISA and OS to any other ISA and OS combination (assuming that it was designed to do so). A slight extension of this abstraction enables virtualization of the ISA, OS, and hardware system, including the Basic Input/Output System (BIOS), I/O, and memory. In this case, the translator becomes an emulator. The Ersatz-11 [1] and the REVIVER-11S [2] are a few examples of software emulating a complete PDP-11 system on your PC under Windows or Linux. Emulators are available for virtually any hardware systems – especially for the older and discontinued ones. Nevertheless, the underlying technology is sound and modern hardware platforms provide enough compute power to be on par and often outperform the original systems.

Of course other abstraction solutions exist, trying to provide the programmers with the "Holy Grail" of Write Once, Run Anywhere. All that said, virtualization is a method to abstract a complete hardware platform, offering to any unmodified software the ability to execute on a fundamentally different hardware platform. If someone uses several of these VMs on the same hardware, virtualization allows the multiplexing of the underlying hardware resources. This brings us back to the 1960s when IBM pioneered virtualization on its mainframes.

9.2 Virtualization and Partitioning

From this point forward, virtualization is assumed to be defined as the very last virtualization approach presented in the preamble: abstraction of a complete hardware platform offering unmodified software the ability to execute on a fundamentally different hardware platform. In addition there are a number of other terms that should be understood in order to fully comprehend virtualization:

- Hardware platform – the sum of the hardware resources that will be multiplexed between applications. Often, the name of a hardware resource is prefixed by `physical` to make the distinction between virtual resources.

- Software abstraction layer – typically executing on the hardware platform, can be either the OS, denoted host OS, or a *Virtual Machine Manager* (VMM), denoted *hypervisor*. The main difference between an OS and a VMM is that the VMM is standalone and is directly providing the virtualization services in addition to some basic OS services (such as scheduling and memory management). In some cases, the VMM is hosted on the OS and relies upon the host OS' services. In both cases, the VMM is creating and managing the VMs.

- VM – hosted by the VMM and capable of executing any unmodified software and OS supported by the VMM. These OSes, termed guest OS, can be different from the host OS. On the top of the guest OS, the unmodified applications are executed as usual. Note as denoted in Figure 9.1, all of the hardware resources are shared between the VMs.

- Partitioning – consists in dedicating the use of a hardware resource to a given VM as denoted in Figure 9.2. This technique is usually employed when performance is a dominant factor. Indeed, if a hardware resource, such as a physical I/O device, is shared between multiple VMs by the VMM, then from

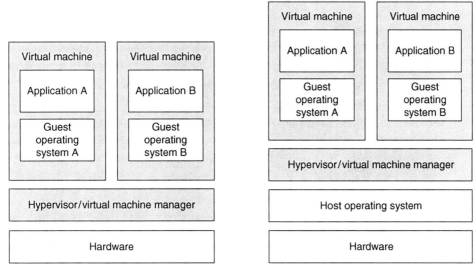

Figure 9.1: Hosted and stand-alone virtual machine manager configurations

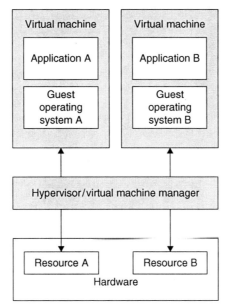

Figure 9.2: Hardware resources partitioning between virtual machines

the point of view of a single VM, the performance of the virtual I/O device (the device exposed to the VM by the VMM) may be suboptimal. A good example of hardware resources that can be partitioned are the processor cores and the Ethernet ports. For example, on a dual-core processor, it is possible to dedicate one core to the first VM, and share the other core between multiple VMs. Similarly, it is possible to assign a couple of Ethernet ports to one VM, and share the other available Ethernet ports between multiple VMs. It is important to notice that virtualization and partitioning are not mutually exclusive, and can both be employed simultaneously. In the last section of this chapter, setup of such a virtualized system will be detailed along with a demonstration of the overall performance that is possible to obtain.

• Para-virtualization – is a virtualization technique which does not allow the execution of an unmodified OS in a VM. The guest OS must be modified to enable it to communicate with the VMM and perform specific calls into it. Obviously, the disadvantage of this approach is the need for a special version of the OS. The benefit is that some interactions with the hardware are more efficient, providing non-negligible performance gains over software virtualization. It also enables the design of simpler VMMs.

9.2.1 VMM Architectures

As mentioned earlier, a VMM can either be standalone or hosted by an OS. The first approach requires the VMM to provide OS-like services and the latter relies upon the host OS's services and its functionality. Although an intermediate solution exists between the hosted and standalone approaches, these two are the commonly employed approaches. It is important to note that neither of these approaches is better than the other; they have advantages and disadvantages which should be taken into consideration when opting for a VMM. To provide a comparison, consider the path taken by an I/O request issued by an application running in a VM. Starting with the hosted VMM approach, the I/O request is sent to the guest OS. The guest OS, via its device driver, attempts to access the hardware. This access is detected and trapped by a small *kernel module*, the VMM monitor, running in conjunction with the host OS. The intercepted request is then transferred to the user-space component of the VMM (running with the same privilege as any other application), which based on the targeted device, converts the I/O request into another I/O request to the host OS. The host OS accesses the physical I/O device via its driver and serves the I/O request. Data is finally sent back to the application using the reverse path. It is possible to shorten the processing path of the I/O request if para-virtualization is employed. In this case, the modified guest OS can directly interact with the host OS to access the hardware. Besides the long path to serve the I/O request, a hosted VMM is dependent upon the host OS's scheduler. More generally it is dependent on the availability and the stability of the host OS, and cannot execute if the host OS is down.

With a standalone VMM, the processing is slightly different. Indeed, the same I/O request is sent to the guest OS by the application. When it is processed via the guest OS' device driver, the request is detected and trapped by the VMM. The VMM then uses its own device driver to access the hardware. The data is sent back to the application using the reverse path. On one hand, the standalone VMM does not depend upon an OS scheduling policy and any security flaws inherent in the host OS. On the other hand, the VMM must have its own driver for the devices it needs to support. Usually, standalone VMMs are credited with a higher level of security because their code is smaller and more controlled than the code of a general purpose OS. However, if a standalone VMM is targeted to support multiple hardware platforms and devices, its code may quickly bloat.

9.2.2 Virtualization's Use Models and Benefits

Virtualization and multi-core processors enable well-established usage models such as those in information technology (IT) shops and end-users. In addition, new non-traditional target verticals such as embedded and telecommunications can now create their own use models, and many of the use models for virtualization have yet to be invented.

In the next sections, an overview of some of the traditional usage models for virtualization starting with the traditional IT shops' environment will be detailed. Second, use models for the telecom and embedded arena are described. Note that even if many of the use models presented here are strongly bound to a server environment, when it is meaningful, examples that also apply to clients will be introduced.

Before diving into details, the following naming convention will be used in the discussions:

- HW – designates a system, a server, a single blade server, or even a desktop client. If multiple systems are involved in an example or a figure, then sequential numbers are used to differentiate them (HW1, HW2, ..., HWn)

- VMMn – virtual machine managers (VMM1, VMM2, …, VMMn)

- VMn – virtual machines (VM1, VM2, …, VMn)

- OSn – operating systems (OS1, OS2, …, OSn)

- Capitalized alphabet letters (A, B, …, Z) denote applications or workloads

By default, no assumptions are made regarding the nature or the configuration of hardware and software components. In other words, OS1 and OS2 are assumed to be different unless otherwise specified. If components are identical, then the same name is used with a prime, second, tierce mark, etc. (A', A", etc.). Finally, if an illustration shows a *before* and an *after* situation, newly introduced components are put in exergue using a gray shading.

9.2.2.1 Workload Isolation

Workload isolation consists in segregating two or more applications from each other. Consider a simple case with applications A and B that are initially running on HW managed by the OS. To isolate A from B using virtualization, a VMM is first installed on HW. Two VMs are created, each of them running its own instance of the operating system (OS' and OS"). One VM will host A, and the other VM will host B. Question: Why should someone isolate workloads? The essential reason is to isolate application failures, rather than the applications themselves. Indeed, in the initial setup, A can interfere with B and vice-versa, even under the scrutiny of a modern OS. In the worst case scenario, A or B can crash the OS itself (i.e., by executing an untested path in a driver, which executes at the highest privilege level, leaving the kernel defenseless). Once A and B are isolated using virtualization, if A shall crash or bring down its OS, B will continue to execute seamlessly. Of course, workload isolation is not the universal panacea as it does not protect against logical application errors. For example, A may bring B down by communicating an out-of-range data value to B that triggers a software bug (a few recent and memorable rocket crashes occurred because software modules exchanged data without converting floating-point data of different precision). Of course, this scenario is symmetrical and can apply to many workloads. Note that workload isolation applies to servers and clients. Figure 9.3 illustrates workload isolation, where the bounding gray box around the OS and applications represents the VM.

9.2.2.2 Workload Consolidation

Workload consolidation consists of reducing the number of physical systems required to run a given set of workloads. Note that *de facto* workload isolation is classified as

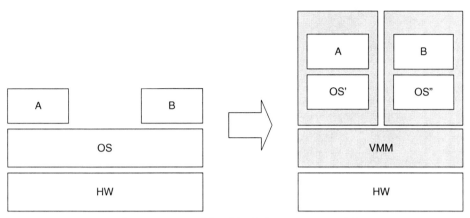

Figure 9.3: Workload isolation usage model

consolidation. This is not mandatory, but definitely makes more sense. By consolidating workloads, it is possible to reduce the computing systems' total cost of ownership (TCO). Indeed, consolidation reduces the number of systems required to run the workloads. From another perspective, workload consolidation increases the compute density, and likely reduces the required power density in a datacenter. To illustrate a workload consolidation usage model, suppose that an IT shop runs three server applications, denoted A, B, and C, on three different hardware and OS stacks. In this initial configuration, the IT department has to maintain and manage three physical systems. Each system requires its own space, its own power source, and uses its fair share of climate control budget in the datacenter. To consolidate workloads A, B, and C, a new system may be required (HW4). HW4 must indeed be dimensioned and provisioned appropriately to sustain the needs of A, B, and C (in terms of computational power, storage, and I/O). This is usually not a problem as hardware refresh cycles introduce more powerful, more power efficient, and less expensive systems. A VMM is then installed on HW4. Three VMs are created, each of them running the initial OS and application stack. Figure 9.4 illustrates workload consolidation.

9.2.2.3 Workload Migration

Workload migration is a very promising usage model of virtualization. It consists of moving a VM and its workload from one virtualized system (HW, VMM stack) to another. Assuming that the virtualized systems exist and are available, workload migration does not require any additional hardware or software resources. By migrating

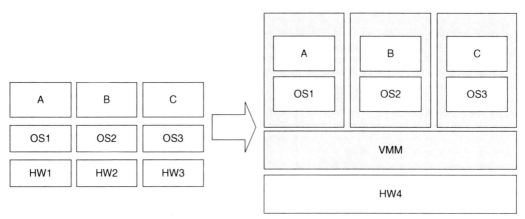

Figure 9.4: Workload consolidation usage model

workloads, it is possible to optimize a data center's utilization rate, its power dissipation, cooling, and much more. In fact, the benefits of workload migration are so numerous and diverse that many usages are still to be discovered. To illustrate the workload migration usage model, suppose there are two consolidated and isolated workloads – A and B – running on one virtualized system denoted (HW1, VMM1). In addition, assume there is a second virtualized system, denoted (HW2, VMM2), in stand-by mode. Although workloads A and B are isolated from each other (workload isolation) and sharing the resources of a unique system (workload consolidation), it may be useful to migrate workload B from the first virtualized system to the second for load balancing purposes. If the load increased substantially for B (i.e., ATMs' back-end processing capacity in Europe were maxed out when countries introduced the Euro currency, removing former national currencies from circulation), it is possible to migrate it to the stand-by virtualized system, providing the required additional computational horsepower to accommodate the load. This operation can also be viewed as workload de-consolidation. Of course, this implies that the target virtualized system is well provisioned. Similarly, once the peak demand period is over, it is possible to re-consolidate the workloads by moving B and its VM back to the initial virtualized system and shutting down the previously enrolled high-performance servers (or suspend it to idle state). Even if the implementation of these few examples showing the use of workload migration requires the definition of elaborated management policies, the possibilities are numerous. One can also use workload migration to transfer a data center's thermal hotspots in the function of the computational

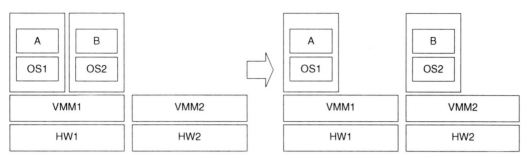

Figure 9.5: Workload migration use model

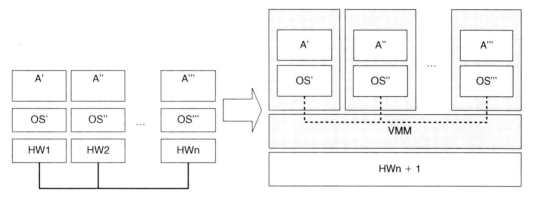

Figure 9.6: Clustering over SMP usage model

load and therefore dynamically optimize the climate control's air flows. Note that it is possible, depending on the VMM used, to perform live migrations. A live migration consists of moving running applications in a VM from one system to another, without having to shut the system down or copy and restart an image of the VM's memory from disk to disk. This usage of workload migration can be part of disaster recovery plans, where services can be restored almost immediately in case of tragedy. Figure 9.5 illustrates the workload migration use model.

9.2.2.4 Clustering Over SMP ... and More

This section discusses a collection of usage models that suit both servers and clients. One common denominator in these usages is the ability to conveniently reproduce eclectic and complex system setups using a single virtualized system. The archetype of this usage model is clustering over symmetric multiprocessing (SMP) as shown in Figure 9.6.

Programming a parallel application targeting a clustered environment is not an easy task. It requires the selection of the right algorithms, the right workload distribution model, and the right communication and synchronization mechanisms. Inherent in this development is the need for debugging. Unfortunately, requisitioning a cluster, even a small one, to debug a program is not the best use one can make of expensive hardware. Alternatively, clustering over SMP, using virtualization, can help. This operation consists first of installing a VMM on an SMP system (i.e., the developer's workstation). The next step consists of creating as many VMs as the virtual cluster size requires. A good practice is to create as many VMs as available processor cores of the virtualized SMP machine. Finally, when the virtual network is configured, it is possible to proceed to the traditional cluster installation. Once the application is debugged and tested, it is still possible to benefit from virtualization by hosting the control node of the cluster in a VM.

Following the same approach, virtualization provides an easy and inexpensive method of reproducing multiple environments for such purposes as software testing and support. A software engineer can create a set of VMs running the major flavors of Linux[*] OS, one VM per major distribution of the Microsoft Windows OS, and even some more exotic OSes, and use them to test a new build of its software. Figure 9.7 depicts such a scenario. Similarly, a support engineer can configure different instances of a reference VM to run various software releases and conduct functional comparisons based on customers' inquiries. Development of embedded applications requiring a cross-platform environment and the debugging of a kernel driver are a few additional examples where virtualization can be valuable.

9.2.2.5 Platform Management

Platform management is a strong proof point for virtualization, especially for the enterprise and large corporations. Although multiple implementation and deployment schemes are possible, the main idea of using virtualization for management and implementing a robust security policy by extension relies on the fact that a VM can run in the background without notice. Therefore, the VM hosting the management software can enforce advanced security policies. It can also provide a suitable administration environment if the IT team has to remotely manage the virtualized system. The VMM itself can be integrated into the platform's firmware (i.e., using the Unified Extensible Firmware Interface (UEFI) infrastructure [3]) and be executed by a baseboard processor, providing a remote security

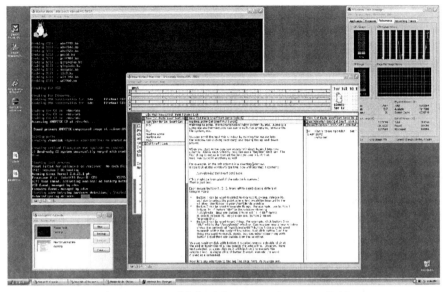

Figure 9.7: Multiple OSes executing over SMP

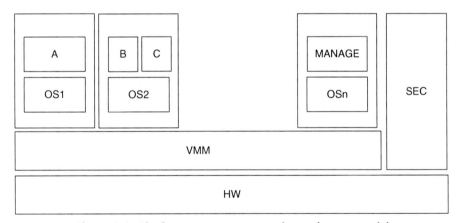

Figure 9.8: Platform management and security use model

and management tool. In Figure 9.8, a simple virtualized system running two VMs is depicted; each VM is hosting OSes and applications. In addition to these VMs, a dedicated VM runs management software and the firmware runs security applications. These two extensions are invisible to the user and the former could even be used offline.

On the client side, virtualization can also improve your system's security level, even with a hosted VMM. Indeed, one can use a VMM and dedicate a VM to browse the Web or evaluate trial software without taking the risk of compromising the underlying system's security. This usage model is gaining in popularity, especially among personal computer users. If the dedicated VM is attacked by a virus or a worm; it suffices to delete the VM's image from the host system's hard drive and replace it with a fresh copy. Undeniably, virtualization offers a good level of protection. Unfortunately, it would be naïve to believe that the host system is out of reach from malicious codes and attacks. Indeed, many research articles are studying VM- and VMM-based attacks demonstrating feasibility of such approaches [4]. In contrast to increasing VMM vendor's awareness about security threats specific to virtualization, the use of a trusted platform as the foundation of a secure VMM remains a plausible answer to this class of threats.

9.2.3 Motivation in Telecom/Embedded

The previously discussed virtualization usage models in a traditional IT environment are certainly valid and fully applicable to telecom and embedded market segment applications; however, there are more specific usage models for virtualization in the telecom and embedded arena that are worth detailing.

9.2.3.1 Business Considerations

Telecommunication applications are subject to very long development and deployment cycles spanning, in some cases, decades. Indeed, typical enterprise applications and platform lifetimes are usually in the range of 5 years. Conversely, telecom and embedded applications can require easily 7–10 years between the definition and the effective mass deployment phases. Therefore, telecom and embedded applications can be deployed as-is, with little or no modifications for decades. Accordingly, it is not uncommon for telecom equipment manufacturers (TEMs) to invest several man-years of research and development into the creation of very specific computational components such as OSes, drivers, middleware, and applications. This specialization effort goes as far as the development of dedicated programming languages, such as Erlang which was created by Ericsson, to satisfy their particular telecom applications' technical requirements. In addition to the software components, TEMs have also developed a broad range of dedicated and specialized hardware to provide the infrastructure for the constantly evolving and demanding communication needs of our modern societies. However, due to cost constraints,

TEMs have undertaken a considerable standardization effort to reduce their development costs, to increase the reusability of their system components, and to improve their solutions' interoperability. In other words, TEMs desire to increase their return on investments and reduce their solutions' time to market. For example, the Advanced Telecom Computing Architecture (ATCA) initiative is one example of such a standardization effort [5].

Although TEMs have embraced the switch to ATCA and the use of commercial off-the-shelf (COTS) server components, they are now facing a non-trivial dilemma: How can they reuse the enormous software stacks they have developed over the decades? Porting their applications to open source OSes and mainstream processor architectures is certainly a meaningful move, but it only makes sense for recently developed applications. It also requires a considerable investment and uselessly increases the time to market of their solutions, which as you recall is the opposite of one of the TEMs' initial objectives. Finally, porting, or rewriting from scratch, legacy, and almost end-of-life software is simply not sensible. Fortunately, virtualization offers a convincing alternative to porting legacy software while allowing the adoption of standard and high-performance hardware.

9.2.3.2 Real-Time OS Support

One of the somewhat unique requirements of telecom and embedded applications is their need for real-time OSes (RTOS – such as the Enea1 OSE* RTOS family or Wind River* VxWorks*). Keep in mind that some of these RTOSes were developed a long time ago by the TEMs, are lightweight, and do not really benefit from recent hardware improvements such as multi-core processors and 64-bit memory addressing. They neither support a broad range of devices. Recall that real-time is not a synonym for `fast`, but simply means that a given operation performed by the OS will always be carried out in the exact same amount of time. It happens that these durations are usually extremely short, but it is not required to earn the qualification of "real-time." It is easy to understand why an embedded or a telecom application requires RTOS support. On one hand, the collection of sensor data of a robot cannot suffer the slightest delay and on the other hand, telecom protocols require the strict respect of timings that solely an RTOS can guarantee. Therefore a VMM suitable for the most taxing telecom and embedded applications must be able to host RTOSes in addition to general purpose OSes. Figure 9.9 shows such usage of virtualization to consolidate multiple workloads, including multiple applications run by RTOSes. Later on, the performance benefits of using such a configuration will be demonstrated.

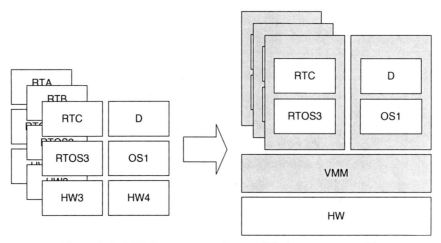

Figure 9.9: RTOS support and consolidation usage models

9.2.3.3 High Availability

Telecom infrastructure equipment is the backbone of our modern communication-centric societies. A common event, such as calling someone using a mobile phone, results in the establishment of a complex control and data path between multiple different computers and network equipment. Thus, between the setup of the radio link between your mobile phone and one of the carrier's base stations and the beginning of your conversation, many systems are involved. Although we can admire the complexity of such an exploit, we rarely forgive the slightest technical issue, such as a dropped call or not being able to contact a relative on New Year's Eve. You probably guess now why *high availability* (HA) [6] is important in the telecom market segment. Recommendation E.800 of the International Telecommunications Union (ITU-T) defines reliability as: "The ability of an item to perform required function under given conditions for a given time interval." The reliability is a probabilistic measure for time interval *t*. In practice though, mean time between failure (MTBF) is used as a measure of reliability. Similarly, availability is defined as: "The ability of an item to be in a state to perform a required function at a given instant of time or at any instant of time within a given time interval, assuming that the external resources, if required, are provided." The availability at instant *t*, is approximated as MTBF/(MTBF+MTTR) – where MTTR is mean time to repair. In other words, reliability (MTBF) refers to failure-free operation during a time interval, while availability (MTBF and MTTR) refers to failure-free operation at a given time.

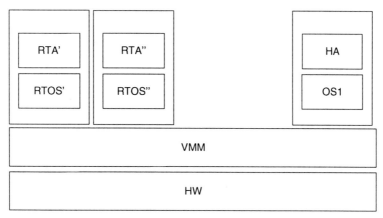

Figure 9.10: HA usage model

HA is typically stated in terms of the average time a system is down, and therefore unavailable to the users. This metric is expressed as a percentage in number of nines. Thus, five nines – one of the highest ratings, corresponds to 99.999% availability. Five nines is equivalent to a qualified system being down for less than 5 minutes a year. It is obvious that all the components of a system (hardware and software) must satisfy a strict list of requirements to earn this rating. In Figure 9.10, a virtualized system is depicted that runs two VMs hosting real-time workloads as well as an HA application, called HA middleware, that ensures the availability and resiliency of the workloads.

9.2.3.4 Break Through the Performance Scaling Barrier

Two traditional approaches exist to host parallel applications and govern the nature of communication. These approaches are tightly linked to the targeted hardware platform's underlying memory architecture. SMP systems are well suited to threaded and multiprocess applications sharing data in memory and communicating via traditional OS infrastructure (shared memory, pipes, etc.). This is the realm of server and desktop applications leveraging multiprocessor and multi-core processor computers. AMP (asymmetric multiprocessing), the second model, prevails in clustered and networked computers. This model naturally puts the emphasis on interconnect, where none of the nodes of the network have direct access to each other's memory. Data exchange is performed through communications via the network, using, for example, a specific API such as Message Passing Interface (MPI). Note that it is possible to have a mixture of both models, where a hybrid compute cluster is built out of multiprocessor and multi-core processor machines. However, with the older applications, code developed to execute on AMP machines is rarely threaded as these hybrid constructs are relatively recent. This is definitely true in the telecom arena, where the development cycles and application lifespans are so long. To seek parallelism, telecom applications rely on the natural parallelism that exists in the very nature of communications: each call is independent and can therefore be handled by different computers. Developers of such applications have naturally adopted the cluster organization where the compute nodes are communicating via a communications backbone and are load balanced by a specialized middleware. Independently of the physical form factor, multiple blade servers can host the same application managed by a middleware, often replicated on two blades for HA purposes, which shares the communication traffic among the blades. Note that little or no data exchange is required between the compute nodes. This is a major difference between applications in this domain and high-performance computing (HPC) applications. Examples of such middleware are TelORB[*] by Ericsson and RTP4[*] by Fusjitsu-Siemens[*].

The problem with this application architecture is that with the use of standardized compute nodes, built upon multi-core processors, very little of the nodes' compute power is used by the applications. As will be demonstrated later in this chapter, virtualization can be advantageously used to break through the non-scaling barrier of legacy telecom applications. The basic idea is to use a VMM to expose as many single processor VMs as processor cores available in the underlying compute blade. Each VM hosts its OS/ RTOS and runs a single instance of a purely serial application. Obviously, the virtualized compute blade must have enough I/O bandwidth to feed each VM. Figure 9.11 depicts this performance scaling use model.

9.3 Techniques and Design Considerations

9.3.1 Virtualization's Challenges on x86 Processors

In this section, essential challenges faced by a VMM designer when targeting x86 processors are discussed. For each challenge, a short explanation is provided as well as how the challenge presents itself in the context of virtualization. The traditional methods of meeting these challenges are discussed followed by their shortcomings in the face of virtualization. Finally, details on how the hardware support provided by the Intel[®] Virtualization Technology supplement traditional methods to meet the challenges.

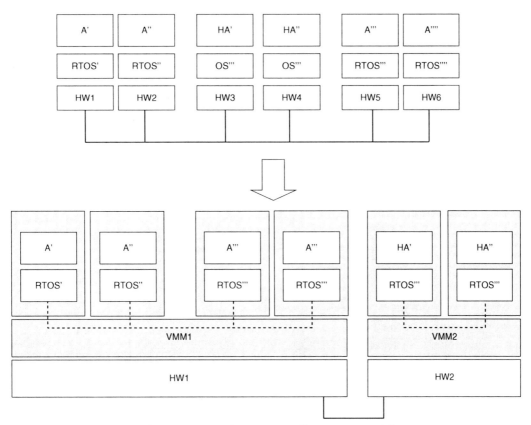

Figure 9.11: Performance scaling usage model

9.3.1.1 Privilege Management

Today, it is expected that applications hosted by a modern OS are isolated from one another. In this setting, if an application crashes, the entire system does not stop functioning; the OS handles the crash and continues execution of the other tasks. This protection mechanism between tasks is enabled by privilege management implemented in modern processors.

On x86 microprocessors, the first two bits of the code segment register (CS register), also known as the current privilege level (CPL) field, are made available to OS developers to facilitate the management of task privileges. Privilege levels are also known as protection rings, or simply rings. Two bits define four possible rings (ring 0 being the most privileged level and ring 3 the least privileged level for a program) as detailed in Figure 9.12.

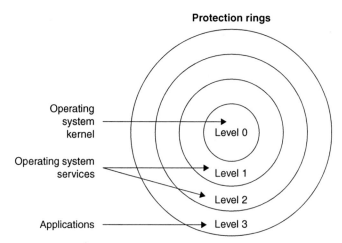

Figure 9.12: Protection rings

The processor core uses the privilege levels to prevent a program executing at a lower privilege level from accessing code, data, or a stack segment of a task executing at a more privileged level. If such an access is attempted, the processor core detects and generates a general-protection exception (#GP – exceptions and interrupts will be detailed later in this section) that the OS handles. In most cases, OSes are using ring 0 to execute their kernel code, ring 1 to run their service routines, and ring 3 to run the user applications; ring 2 is not typically used. Based on the previous description, any task can read its own CPL very simply as shown by the register window of a debugger in Figure 9.13. Later it will be shown why this simple and anodyne fact can be problematic for a VMM designer.

Microprocessors expose several registers to facilitate the implementation of an OS which can be problematic for correct implementation of virtualization. For example, the Global Descriptors Table Register (GDTR) is a critical OS register that the privilege mechanism prevents modification from ring 3 applications. Therefore, if the code depicted in Figure 9.14 is executed by a ring 3 application where a value is loaded into the GDTR by using the LGDT instruction, an exception is generated which is trapped by the OS. Unfortunately, this register and others can be read which causes a problem that will be detailed later in the chapter. Similarly, any user task (running in ring 3) can use the SGDT instruction to store the content of GDTR into a memory location (as soon as it is at least 6 B long). These two examples show that a user application can monitor some of the essential activities of an OS by simply reading a few registers. And of course, the OS

```
Registers                                                    ✕
    EAX = CCCCCCCC EBX = 7FFDC000                            ∧
    ECX = 00000000 EDX = 00000001
    ESI = 00000040 EDI = 0012FEDC
    EIP = 00411A32 ESP = 0012FE00
    EBP = 0012FEDC EFL = 00000206


    CS = 001B DS = 0023 ES = 0023 SS = 0023
    FS = 003B GS = 0000


    0012FED0 =                                               ∨
  🖳 Autos │ 🖳 Locals │ ▤ Output │ 🖳 Watch 1 │ ᵒˣ Registers
```

Figure 9.13: CS register reveals current privilege level

```
void main(void) {
      char zone[6];
      _asm {
            sgdt zone
            lgdt zone
      }
}
```

Figure 9.14: Code example accessing GDTR

itself can do it, as well as modify the content of those registers. So why is it a problem for a VMM designer?

To understand the problem, consider the scope of a virtualized environment where a VMM controls one or more VMs. Each VM hosts an OS, which in turn is responsible for executing various user applications. Obviously, the situation from the point of view of a user-level application does not really change with respect to privilege management and these programs will continue to be able to read some exotic processor registers and will be irremediably intercepted as soon as they try to modify any of them. From the point of view of the VMM, the situation is also clear as it plays a role similar to the OS in a non-virtualized environment. Indeed, the VMM will control the processor and will advantageously use the privilege mechanism to protect each VM from the other VMs. You have probably guessed it by now; the issue is with the OS itself. These guest OSes have been bumped out from ring 0 and live in a less privileged level than the one for which they were designed. Unfortunately, it cannot be assumed that the OS will blindly believe that it executes in ring 0. For example, the Linux OS employs the function, `user_mode()`, which is defined in `/include/asm-i386/ptrace.h` to test if the current code is executing in ring 3. One could think about using ring 2 to run the guest OSes, but it is not a good solution for different technical reasons. Nonetheless, it is also not reasonable to hope that the industry will modify all existing OS code for the sake of virtualization. Therefore, ring 3 is usually the privilege level used in pure software implementation of a VM, with the VMM being responsible to fool the OSes into believing that they run in ring 0. This problem is also called *ring aliasing* in the literature.

As you may guess, the illusion not only requires intricate software techniques, but also has a consequence of negative performance impact on the virtualized system. A good example of such adverse performance impact is the SYSENTER and SYSEXIT instruction pair. These instructions were introduced with the Intel Pentium® II processors to provide a low latency, highly optimized system call mechanism to switch between ring 0 and ring 3. For example, these instructions allow a user application to call system level routines very efficiently.

SYSENTER will switch the processor into ring 0 and SYSEXIT will switch the processor from ring 0 into ring 3. Here, the performance gain comes essentially from the fact that SYSENTER is configuring the processor by using a fixed scheme as depicted in Figure 9.15. For the variable part which corresponds to the location in the target

```
CS.SEL := SYSENTER_CS_MSR // Operating system provides CS

// Set rest of CS to a fixed value
CS.SEL.CPL := 0 // CPL   0
CS.SEL.BASE := 0 // Flat segment
CS.SEL.LIMIT := 0xFFFF // 4G limit
CS.SEL.G := 1 // 4 KB granularity
CS.SEL.S := 1
CS.SEL.TYPE_xCRA := 1011 // Execute + Read, Accessed
CS.SEL.D := 1 // 32 bit code
CS.SEL.DPL := 0
CS.SEL.RPL := 0
CS.SEL.P := 1
```

Figure 9.15: SYSENTER call semantics

code segment where to jump on SYSENTER execution, the processor is using an MSR configured by the OS at boot time.

As stated previously, SYSEXIT will switch the processor from ring 0 into ring 3. In fact, it follows a mechanism similar to the one used by SYSENTER. The main difference is that the return address in user space is not provided by an MSR (which would not be a viable solution anyway), but is maintained by the caller (before performing the SYSENTER – via the EDX and ECX registers). With a virtualized environment, if a user application calls a system routine via SYSENTER (actually via a library call to access, e.g., to a file or allocate some memory), then control will be given to the VMM – assuming that it configured the MSRs correctly – and not the host OS as expected. It is then the responsibility of the VMM to ensure that the illusion is maintained. With SYSEXIT, there is an additional risk as the instruction itself will check its original privilege level and if it is not ring 0 then it will fail as depicted in Figure 9.16.

To work around the ring aliasing problem, the processor support for virtualization makes available to VMM designers a new set of operations called VMX operations. These operations exist in two flavors: root and non-root. By analogy to the UNIX OS's root super user, one can consider the non-root mode as being less privileged than the root mode. Naturally, the VMM operates in root mode, while the VMs and their guest OSes are executing in non-root mode. Transitions between both modes are known as VMX transitions. Switching from root to non-root mode is called a VM Entry transition (control is transferred from the VMM to the VM) while the opposite is called a VM Exit transition (control is transferred from the VM to the VMM). Figure 9.17 depicts these transitions and Figure 9.18 illustrates the transitions over time.

There are two essential differences between the root and the non-root mode. First, the new instructions introduced with Intel® VT-x[1] are only available in root mode. If the

```
SYSEXIT
IF SYSENTER_CS_MSR == 0 THEN #GP(0)
IF CR0.PE == 0 THEN #GP(0)
IF CPL <> 0 THEN #GP(0)
...
ESP := ECX
EIP := EDX
```
Figure 9.16: SYSEXIT ring 0 verification

[1] 10 op-codes are added (see Tables 9.2 and 9.3).

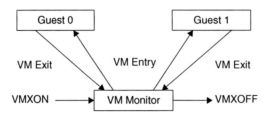

Figure 9.17: VMM and guests interactions

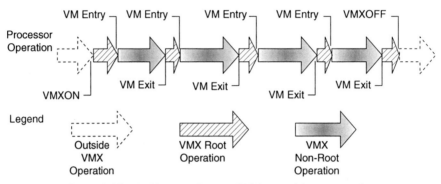

Figure 9.18: VMX operation and VM transitions over time

instructions are executed in non-root mode then a VM Exit transition is initiated. Second, the behavior of the processor is slightly altered in non-root mode in such a way that some privileged operations initiate a VM Exit transition, hence, giving control to the VMM for proper resolution of the altered privileged instructions. Yet, it is very important to understand that the root or non-root mode doesn't change the role of the CS register's CPL field. Therefore, a guest OS executing in a VM (in non-root mode) is executing as expected in ring 0 and is therefore unable to detect the presence of the VMM. From a performance point of view, the objective for a VMM designer is to keep the execution of the VM as long as possible in non-root mode.

9.3.1.2 Memory Management

Memory management is a complex topic and most of its details are out of the scope of this chapter. However, to understand the address space-related issues raised by virtualization, understanding some basic techniques is crucial. For an x86 processor, accessing memory requires translation between addresses issued by the applications,

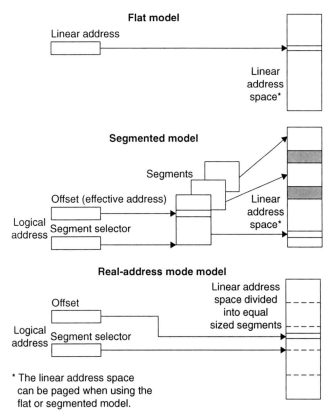

Figure 9.19: IA-32 Architecture memory management models

including the OS, and addresses that the physical memory system can understand and use to read and write (see Figure 9.19 – showing two other memory management modes, not covered in this chapter). The principal problem is to convert a virtual address composed of a 16-bit segment selector and a 32-bit offset in the selected segment into a linear, physical address. As a tip, the simplest mental representation you can build when thinking about physical memory is of a huge vector of bytes, where each byte has a unique address.

Two distinct, but complementary, mechanisms are used to achieve this goal, segmentation and pagination. Segmentation converts the virtual addresses into linear virtual addresses by splitting memory into segments which are fully qualified by its base address, its limit, and its attributes. While the virtual address space is huge (i.e., 64 TB for an Intel® 386 processor), the linear virtual address space is limited to 2^{32} bytes on a 32-bit processor (or 2^{64} bytes on 64-bit processors). This linear virtual address is then translated via pagination into a physical address. Segmentation and pagination are performed using two in-memory tables managed by the OS. While the segment table is in the linear address space, the page table is in the physical address space. This organization allows these two mechanisms to operate simultaneously without interfering with each other. In fact, segmentation uses two tables: the Global and Local Descriptors Table (GDT and LDT), each of them holding segment descriptors. A given segment descriptor is identified via a descriptor table identifier and an index into that table. Note that the virtual address space is split between the GDT and the LDT, and during a context switch, the GDT stays unchanged, while the LDT is updated. By having a unique LDT per task – although it is not mandatory – the OS can enforce protection between tasks.

Each segment descriptor is bound to a segment selector. The 16-bit segment selector holds the two bits of the Requested Privilege Level (RPL) field, the one-bit Table Index (TI – indicating the descriptor table to use, GDT or LDT) flag, and the 13-bit index into the adequate descriptor table. The net result is that a segment selector and an offset (the virtual address), along with the segment descriptor's base address in the linear address space, can fully qualify a byte in any valid segment (stack, data, or code) as soon as the CPL of the running task is lower (more privileged) than the segment's RPL.

The GDT Register (GDTR) of the processor holds the linear base address of the GDT, while the LDT Register (LDTR) holds the linear base address of the LDT as depicted in Figure 9.20. These registers are loaded and stored using the LGDT, SGDT, LLDT, and SLDT instructions. Note that the segment holding the LDT modified via LLDT or SLDT must have an entry in the GDT.

Whereas segmentation splits the virtual address space into virtually any sized segments, pagination divides the linear and the physical address space into equally sized blocks, called pages that are typically 4 KB in size and 4 KB aligned. Any page from the linear address space can be mapped to any page of the physical address space. Pagination translates the 20 least significant bits (LSBs) of the virtual address into the 20 most significant bits (MSBs) of the physical address. An invalid page access can be caused by the OS restricting access to the page or because the page is in virtual memory and

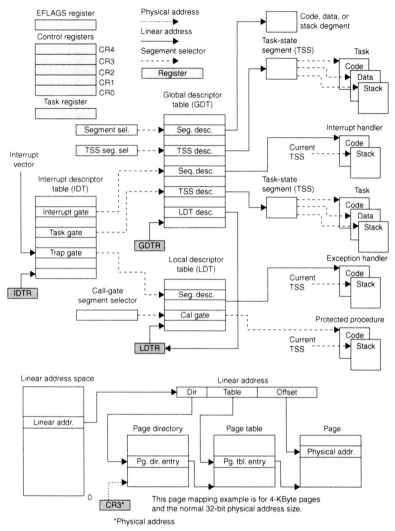

Figure 9.20: IA-32 Architecture system-level registers and data structures

is trapped and processed accordingly by the OS. To accomplish this, a page table is used and maintained in the physical address space. With 4 KB pages and using a 32-bit addressing mode, the 12 LSBs of a page's linear address are null. Therefore, the page table should only have 2^{20} entries (of 4 bytes each, which represents 4 MB of data).

To save memory, the page table is stored in two levels, each of them handling 10 bits of the 20 MSBs to be computed. The first level is called the page directory (PD – a single 4 KB page holding 1 KB of 4 byte entries) and is pointed to by the CR3 register of the processor. CR3 is one of the many control registers of an x86 processor. As the name suggests, these registers control the behavior of the processor.

Each PD entry contains the base address of a second-level page table. The correct entry of the PD is selected by the 10 MSBs of the linear address. The next 10 bits of the linear address (21:12) are used as an offset into the second-level page table. This last entry holds the base address of a page in the physical memory. To finish the translation, the 20 MSBs of the base address in the physical memory are combined with the 12 LSBs of the linear address. Note that the translation is performed by the processor's Translation Look-aside Buffer (TLB), fed with data from the page table hierarchy.

If you should remember one fact from the previous description it is that some of the processor's registers and the associated in-memory data structures are vital to any OS for its memory management. Therefore, memory management is potentially an issue for a VMM designer. Indeed, the guest OS assumes that it has access to the entire address space and that it is in charge of managing it in the way we described earlier. In the same way, a VMM is in charge of managing the virtualized system's address spaces. In addition, because the VMM needs some memory to store its code, data, and stack, it is inevitable that conflicts may arise between the VMM and the guest OSes. Even though the VMM would use a completely different virtual address space, it still must keep in the linear address space some of its essential data structures for memory management such as the GDT, and the *Interrupt Descriptor Table* (IDT). Without hardware support, the VMM must detect each access by the VMs to the linear address space used by the VMM and emulate them appropriately to ensure the correct execution of the VMM and VMs. This issue is also known in the literature as *address-space compression*.

To solve address-space compression, Intel® VT-x uses an essential data structure called the Virtual Machine Control Structure (VMCS). VMCS is designed to manage and ease transitions between the VMM and its VMs. A VMM can employ the VMCLEAR, VMPTRLD, VMREAD, and VMWRITE instructions to control the VMCS for each VM it is managing (or multiple VMCS per VM if it is an SMP). Although the size of a VMCS is processor specific, the memory region associated with a VMCS is limited to 4 KB. VMCS regions reside in the processor's physical address space, in writeback cacheable memory for best performance, and are pointed to by VMCS pointers (64 bits wide address, 4 KB aligned). To make a VMCS active, the VMM uses the VMPTRLD instruction to load the VMCS' address into the processor's current-VMCS pointer. Subsequent operations involving a VMCS will apply to the current VMCS.

Without going into extreme details, there are six zones defined in a VMCS and these are summarized in Table 9.1. The two essential zones of a VMCS are the Guest and Host State Areas (GSA and HSA). GSA is used to save the processor's state on a VM Exit transition and to load the processor's state on a VM Entry transition. Similarly, HSA is used to load the processor's state on VM Exit transition. These two areas facilitate the swapping of the processor's states on transitions between root and non-root modes. The GSA stores the state of CR0, CR3, CR4, Debug Register 7 (DR7 – which play a special role of debug control register), RSP, RIP, and RFLAGS, and for the CS, SS, DS, ES, FS, GS, LDTR and TR registers, their selector, base address, segment limit, and access rights. Note the use and role of most of these components have already been explained. The stack, instruction pointers, and the flags registers are prefixed by an R – which means that they are considered in their 64-bit version in the VMSC. In addition to these essential registers, the GSA stores other processor's states that are not accessible to any software via a register. This is not only helpful in root to non-root transitions, but is also a performance enhancer as some of these shadow registers/regions are kept active between transitions. Saved regions include the Activity State (running, halted, shutdown, waiting on Startup-IPI, etc.), Interruptibility State (temporally blocked events), Pending Debug Exceptions (deferred exceptions), and the Reserved for Future Use VMCS Link Pointer.

Table 9.1: VMCS zones' role description

VMCS Zone	Role
Guest state area	Processor state is saved into the guest-state area on VM Exits and loaded from there on VM Entries.
Host state area	Processor state is loaded from the host-state area on VM Exits.
VM execution control fields	These fields control processor behavior in VMX non-root operation. They determine in part the causes of VM exits
VM exit control fields	These fields control VM exits.
VM entry control fields	These fields control VM entries.
VM exit information fields	These fields receive information on VM exits and describe the cause and the nature of VM exits. They are read-only.

Table 9.2: VMCS maintenance instructions

Instruction	Role
VMPTRLD	Takes a single 64-bit source operand in memory. It makes the referenced VMCS active and current.
VMPTRST	Takes a single 64-bit destination operand that is in memory. Current-VMCS pointer is stored into the destination operand.
VMCLEAR	Takes a single 64-bit operand in memory. The instruction sets the launch state of the VMCS referenced by the operand to "clear," renders that VMCS inactive, and ensures that data for the VMCS have been written to the VMCS-data area in the referenced VMCS region.
VMREAD	Reads a component from the VMCS (the encoding of that field is given in a register operand) and stores it into a destination operand.
VMWRITE	Writes a component to the VMCS (the encoding of that field is given in a register operand) from a source operand.

The register-state composition of the HSA is close to the GSA's, with the following differences:

- There is no field for the LDTR selector, but there is one for the GDTR and the IDTR.

- The IA32_SYSENTER_CS and IA32_SYSENTER_ESP MSRs are stored.

Reconsider the non-faulting instructions that caused a potential threat to the VMM designer. On a VM Exit transition, the processor uses the VMCS to provide crucial information to the VMM on why the transition to root mode has occurred. The VMM can now take the appropriate action based on the transition cause. Tables 9.2 and 9.3 provide a summary of VMX instructions [7].

9.3.1.3 Interrupts and Exceptions Management

The last important topics to cover in this section, especially when you are considering the embedded and the telecom arena are interrupts and exceptions. In the following discussion, assume the generic term event designates both interrupts and exceptions when there is no reason for making a distinction. On one hand, interrupts are delivered to the processor's core only once and are handled in a predefined way (this operation is usually called servicing the interrupts). Interrupts are asynchronous and are generated by external causes (i.e., I/O, timers, sensors, etc. and are not related to the code that the

Table 9.3: VMX management instructions

Instruction	Role
VMCALL	Allows a guest in VMX non-root operation to call the VMM for service. A VM exit occurs, transferring control to the VMM.
VMLAUNCH	Launches a virtual machine managed by the VMCS. A VM entry occurs, transferring control to the VM.
VMRESUME	Resumes a virtual machine managed by the VMCS. A VM entry occurs, transferring control to the VM.
VMXOFF	Causes the processor to leave VMX operation.
VMXON	Takes a single 64-bit source operand in memory. It causes a logical processor to enter VMX root operation and to use the memory referenced by the operand to support VMX operation.

processor is currently executing except if it is issuing an INT instruction). On the other hand, exceptions are synchronous and are used to signal unusual conditions (division by zero, illegal memory operand, etc.) detected while the processor is executing instructions from a program. Interrupts and exceptions usually occur between two instructions at least when they are allowed, or not masked. Also note that some instructions, such as CMPS, CMPSB, etc., used with the Repeat String Operation prefix, may generate an exception between iterations of an instruction.

The easiest method of representing the event handling mechanism is to imagine that a control transfer is performed between the program which was executed by the processor when the event happened, and the routine responsible for servicing the event. After completion, the event servicing routine gives control back to the interrupted program. However, rather than using a CALL instruction (as it is usually performed to transfer control in a program), a special mechanism is used by the processor to perform the non-existing CALL instruction. At the end of the event service routine, an IRET instruction gives control back to the interrupted program at the very location where the processor was interrupted as depicted in Figure 9.21. To do so, IRET reads the IP, CS, and EFLAGS registers from the stack, where these values are placed before the control transfer.

Events are associated with a unique 8-bit value (0–255). There are 18 predefined events for an x86 processor and 224 re-definable and maskable interrupts (from 32 to 255 – the

```
RETURN-TO-SAME-PRIVILEGE-LEVEL: (* PE=1, VM=0 in flags image, RPL=CPL *)
IF EIP is not within code segment limits THEN #GP(0); FI;
EIP <- tempEIP;
CS <- tempCS; (* segment descriptor information also loaded *)
EFLAGS (CF, PF, AF, ZF, SF, TF, DF, OF, NT) <- tempEFLAGS;
```
Figure 9.21: IRET operations (assumes return to same ring)

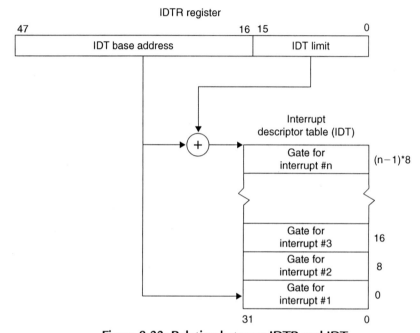

Figure 9.22: Relation between IDTR and IDT

14 missing values are reserved for future usage). This unique 8-bit value is used in conjunction with a table stored in the processor's linear address space, the IDT. The linear base address of this table is stored in the processor's IDT Register. The contents of the IDTR are depicted in Figure 9.22. The *interrupt vector* is the offset into the IDT corresponding to an interrupt and is the unique 8-bit value introduced earlier. The IDTR holds the IDT's base address and also a 16-bit wide table limit. The IDTR is loaded and stored via the LIDT and SIDT instructions. Note that the SIDT instruction can be executed from ring 3, whereas LIDT can be executed only from ring 0. The IDT itself has

a similar structure to the GDT and LDT presented in the memory management section. Each IDT entry is a call gate descriptor to the event servicing procedure. A call gate descriptor is 8 bytes in length and contains a pointer (48 bits wide – selector and offset) and 16 bits of attributes. The selector points into the GDT or LDT, and must select a code segment. The offset of the call gate descriptor gives the entry point for the event handler in the selected code segment. However, there are some differences in the transfer mechanism depending upon if the service routine resides in the same code segment as the interrupted program (i.e., if it is an exception trap) or different segments with different privilege levels. Readers interested in more details can read the Intel® 64 and IA-32 Architectures Software Developer's Manual: System Programming Guide available on Intel's website.

Managing events, especially interrupts, requires special care from a VMM designer. Similarly to the memory aliasing support, Intel® VT-x helps solve the interrupt virtualization problem by allowing the VMM and the VMs to manage their own interrupt and exception control structures (the IDT and IDTR) and to define their own interrupt vectors (host and guest vectors, that may or may not overlap). As explained earlier, there are 32 predefined exceptions defined by the x86 architecture. Exceptions are synchronous events and thus will be generated in the VM that encounters the unusual condition. The problem is that the VMM may require being informed of the occurrence of such an exception or even process it. For example, a page fault is something that the VMM should track, whereas a division by zero can be handled by the VM. To allow such fine-grain handling of exceptions, the VMCS has a 32-bit wide bitmap called the exception bitmap (EB). Each bit set to 1 by the VMM will force the VM to exit, via a VM exit transition, and give control to the VMM when the mapped exception is detected in the VM. The VMM will then use its own IDT to service the exception, or at least account for it. Note that the VMM has access through additional fields in the VMCS to information on the trapped exception. Similarly, if an exception mapped to a cleared bit in the EB occurs, then the VM will not exit and it will use its own IDT to handle the exception.

The remaining 224 external interrupts are supported by Intel® VT-x in a different and richer way. This additional flexibility is required, as there are multiple models to manage hardware interrupts. If the VMM is solely responsible for servicing an interrupt, then the traditional mechanism described previously in this section is used. However, if the VMM expects a VM to process the interrupt too, then it can do so by using a virtual device to generate a virtual interrupt, different from the original vector, to the VMM. If servicing an interrupt is the responsibility of a given VM then the VMM can deliver it to the VM through a host vector to guest vector mapping. This mapping uses the interrupt injection mechanism of Intel® VT-x controlled by the VMCS to communicate the interrupt vector to the VM. Then, during the next VM Entry transition, the processor will use that information in conjunction with the guest IDT to call the interrupt service routine.

Up to this point, it was assumed that the VMM was executing while external interrupts occur. So the last question to answer is what is happening if the interrupts occur while a VM is executing? Once again, the VMX transition mechanism associated with the VMCS provides the answer thanks to the Interrupt Window Exiting bit in the VM Execution Control Fields field of the VMCS. When the processor's Interrupt Flag (RFLAGS.IF – which controls if the processor handles maskable hardware interrupts) is set and there are no other blocking mechanisms impeding a non-maskable hardware interrupt delivery, then the VM gives control to the VMM via a VM Exit transition. This same mechanism allows the VMM to queue an interrupt to a VM while this one is in an interruptible mode. For example, if the VM sets IF using the Set Interrupts (STI) instruction then the VM will be interruptible again (and the interrupt window exiting control was set by the VMM), a VM Exit transition will take place giving back control to the VMM. The VMM knows that the VM is interruptible again, and can deliver an interrupt using one of the methods described earlier.

Of course, we are just scratching the surface of interrupt processing, but the description suffices to illustrate the point. More information is available in *Intel® 64 and IA-32 Architectures Software Developer's Manual Volume 3B: System Programming Guide, Part 2* available on Intel's Website. Figure 9.23 depicts the various paths to process external interrupts by the VMM and its VMs.

9.3.1.4 Missing Support and Future Roadmap

The attentive reader would certainly notice that what has been described up to now is only what has been announced in the beginning of this section: processor virtualization. Although processors are essential components of a computer system, they are not sufficient to build a functional ensemble. One of the missing components, essential to telecom and embedded systems, is I/O virtualization hardware support, especially for the network I/O. Figure 9.24, through the Intel Virtualization Technology evolution multi-generational roadmap, shows that one of the next virtualization enhancements to be expected is precisely the I/O device virtualization (a.k.a. Intel® VT-d). Until this

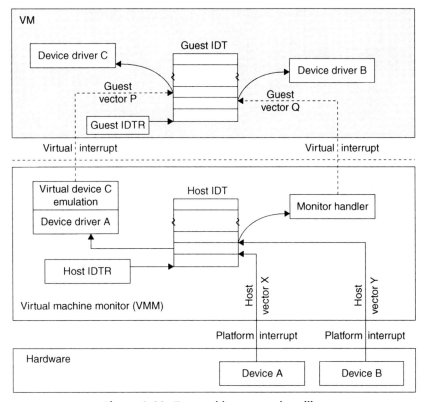

Figure 9.23: External interrupts handling

technology is available, VMMs will use pure software virtualization techniques to share I/O devices. If performance is critical, partitioning will be employed.

It is important to mention that the entire industry is working on I/O virtualization. A remarkable initiative is currently being driven by the Special Interest Group responsible for the PCI Express® industry-standard I/O technology (PCI-SIG®) and is called the PCI-SIG I/O Virtualization specifications [8]. The specifications are aimed to allow concurrently executing OSes above a single computer to natively share PCI Express® devices. These specifications address three main areas:

1. Address Translation Services (ATS) to use translated addresses in support of native I/O Virtualization for PCI Express components.

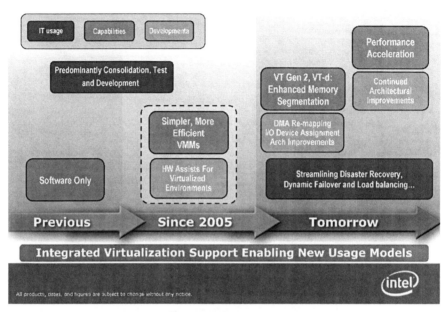

Figure 9.24: Intel® Virtualization Technology evolution

2. Single Root IOV to provide native I/O Virtualization in existing PCI Express topologies with a single root complex.

3. Multi-Root IOV, relying upon Single Root IOV to provide native I/O Virtualization in new topologies (such as blade servers) where multiple root complexes share a PCI Express hierarchy.

9.4 Telecom Use Case of Virtualization

In this final section, the practical side of virtualization is detailed by discussing the use of Intel® Virtualization Technology in the configuration and the performance evaluation of a virtualized ATCA multiprocessor multi-core single board computer (SBC) with VirtualLogix™ VLX for the Network Infrastructure virtualization solution. In particular, the feasibility and benefits of using virtualization to leverage state of the art multi-core architecture and scale legacy applications in the telecom arena are detailed. Figure 9.25 depicts several SBCs mounted in racks typical of use in a production environment.

Figure 9.25: Telecom applications hosted by ATCA SBCs

9.4.1 Setup and Configuration BKMs

VirtualLogix™ VLX for the Network Infrastructure virtualization solution (VLX) is designed to meet the embedded TEMS' and software developers' needs. Therefore, VLX provides full control over how virtualization and partitioning are performed. On one hand, this imposes a few additional steps in the virtualized system's setup compared to off-the-shelf virtualization software; however, it provides full control over the hardware. VLX can be installed on the developer's cross-platform workstation by following the installation directions; a screenshot of the installation is depicted in Figure 9.26.

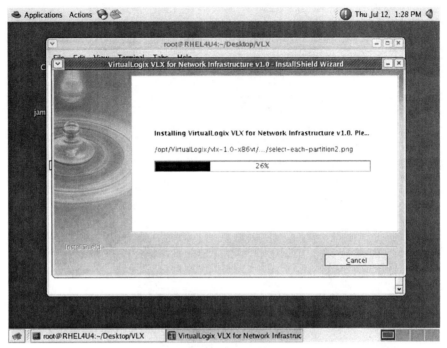

Figure 9.26: VirtualLogix VLX for network infrastructure installation

To demonstrate how flexible VLX can be, the Intel NetStructure® MPCBL0040
SBC [9] as depicted in Figure 9.27 is used to host two VMs, running the Linux OSes
and controlling each, via the native Linux network drivers, and two GbE ports. This
configuration will be partially reused later in this section during a discussion of the VMs
compute and I/O performance.

The first question to answer is how communication between the VMs will be handled.
SBCs are typically headless systems that require a serial port to communicate with the
external world using a terminal. For this purpose, VLX offers a virtualized Universal
Asynchronous Receiver/Transmitter (UART i8250). This virtual device will therefore be
available to each OS hosted by the VMs. Note that VLX also offers a virtualized interrupt
controller (i8259 PIC, Local & I/O APIC), timer (I8254 timer), and a clock (MC146818
RTC). Under Linux, using a boot option such as `console=ttyS0,9600n8`
configures the serial port at 9600 bps, no parity and 8-bit data (e.g., see Figure 9.28).

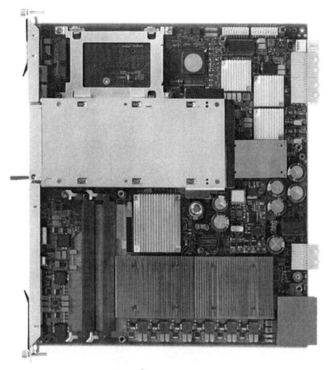

Figure 9.27: Intel NetStructure® MPCBL0040 SBC (ATCA form factor)

These values should be set to the SBC BIOS's default or current settings. Another setting you might want to control is console redirection to the serial port on the front panel. In the MPCBL0040's BIOS, the option is named Redirection after BIOS POST, and should be set to `Always`.

A second crucial question to answer is where the OS and application files will be stored. The MPCBL0040 SBC has support for one Serial Attached SCSI (SAS) hard drive. Although it is possible to add a second SAS drive using a Rear Transition Module (RTM) as depicted in Figure 9.29 or using the SAS ports on the front panel (two SAS ports, one physical SAS connector), the system will be configured with a unique SAS drive to be shared between the VMs. The rational behind this choice is that the applications will not generate intense disk I/O during execution. Note that most telecom applications deployed in the field are running on diskless SBCs and are loading a boot image from a Trivial File Transfer Protocol (TFTP) server via the Preboot Execution Environment (PXE)

Figure 9.28: COM port configuration dialog

protocol. Therefore, the first VM will drive the SAS drive and will export the second VM's file system via virtual block device (VBD) drivers. Both VMs will communicate through a virtual Ethernet segment. Once these issues are addressed, the headless OS installation (redirecting the console to the serial port using the previously set options) can proceed. The only part of the Linux OS installation that will require special attention is the partitioning of the attached SAS drive. Indeed, partitions (at least root and swap) for both VMs will need to be created. Therefore, you will have to go with a manual expert level partitioning as depicted in Figure 9.30. Choose and mount one root and one swap partition for the installation. Remember that the OS image will be used later to create the VMs' boot images by VLX.

The rest of the VM creation process can be accomplished using the command line interface. VLX includes a set of ready-to-use shell scripts which simplifies the

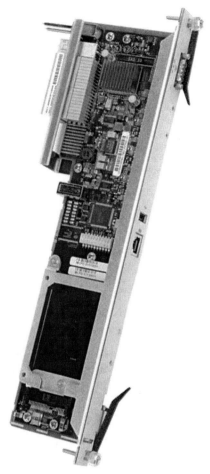

Figure 9.29: Rear transition module for the Intel NetStructure® MPCBL0040 SBC

configuration of VMs and creation of OS boot images. Detailing all of the configuration phases are out of the scope of this chapter; however, a subset of the core configuration file as it controls virtualization and partitioning of the hardware will be detailed. But before going any further, it is highly recommended that you have the technical reference manuals of your target platforms. In this case study, the manual can be found at: http://www.intel.com/design/network/manuals/309129.htm

```
LABEL=/                /              ext3    defaults                    1 1
LABEL=/L2              /L2            ext3    defaults                    1 2
LABEL=/L3              /L3            ext3    defaults                    1 2
LABEL=/L4              /L4            ext3    defaults                    1 2
none                  /dev/pts       devpts  gid=5,mode=620              0 0
none                  /dev/shm       tmpfs   defaults                    0 0
none                  /proc          proc    defaults                    0 0
none                  /sys           sysfs   defaults                    0 0
LABEL=SWAP-sda6       swap           swap    defaults                    0 0
LABEL=SWAP-sda7       swap           swap    defaults                    0 0
LABEL=SWAP-sda8       swap           swap    defaults                    0 0
LABEL=SWAP-sda9       swap           swap    defaults                    0 0
/dev/scd0             /media/cdrom   auto    pamconsole,exec,noauto,managed
0 0
```

Figure 9.30: SAS drive partitioning hosting up to four VMs

Figure 9.31 shows the section of the configuration file of interest. The first VM, labeled Linux_1, is defined between lines 1 and 21. The second VM, labeled Linux_4, is defined between lines 23 and 37. First, Linux_1 will be detailed and then Linux_4's configuration will be commented upon. At the end of the review, all the options will be covered and explained.

The first option (line 2) defines which processor the VMM should use to run Linux_1. In this case, processor 0 is core 0 of processor 0 because the MPCBL0040 SBC has two Dual-Core Intel® Xeon® LV 2.0 GHz processors. Similarly line 23 assigns processor 3 (core 1 of processor 1) to Linux_4. The second option (line 3) defines the behavior of the VMM when a hardware reset is initiated. As it is defined, it will reset all of the VMs. By default, which makes sense in a production environment, only the guest who issued the board reset will be reset. Line 4 defines the method used to manage and allocate memory. Linux_1 will use half of the available memory (vram=50%), and Linux_4 will use the other half (line 26). The vram-native directive enables a one-to-one memory mapping to the guest physical space. In this case, the guest OS image must be linked on the memory bank address (VLX takes care of this when building the OS images). In addition, this option ensures that the VM has access to the first megabyte of the physical memory (containing for example the VGA and I/O memory). Finally, the vdma-enable directive activates direct memory access (DMA) on the PCI bus, allowing the guest OS kernel to set the master enabled bit in the command register of the PCI devices owned by the VM (and used consequently by the PCI bus master DMA). Line 5 defines the I/O ports and the PCI devices that may be accessed by the VM. By using the wildcard character (*) for both vio and vpci entries of the configuration file, Linux_1 is given control over all I/O ports and PCI devices. Note that by default, no access is granted to

```
001:  linux_1="\
002:   hcpus=(0) \
003:   guest-hard-reset \
004:   vram=50% vram-native vdma-enable \
005:   vio=(*) vpci=(*) \
006:   virq=(*) \
007:   virq=(17) \
008:   vpci=(3,0,1) \
009:   vpci=(4,0,1) \
010:   vpci=!(3,0,0) \
011:   vpci=!(4,0,0) \
012:   vpci=!(2,0,0) \
013:   vpci=!(2,0,1) \
014:   pirq=+(9-11) pirq=+(5-6) \
015:   vdevs-size=32k \
016:   vtsc-offset \
017:   vmac=(0;0:0:0:0:0:1) \
018:   vdisk=(3,3,1/8,3,rw) vdisk=(3,3,2/8,5,rw) \
019:   vdisk=(4,3,1/8,6,rw) vdisk=(4,3,2/8,7,rw) \
020:   vdisk=(5,3,1/8,8,rw) vdisk=(5,3,2/8,9,rw) \
021:   root=/dev/sda1 noirqdebug acpi=off console=ttyS0,9600 guest-initrd"
022:
023:  linux_4="\
024:   hcpus=(3) \
025:   guest-hard-reset \
026:   vram=50% vram-native vdma-enable \
027:   vpci=(2,0,0) \
028:   vpci=(2,0,1) \
029:   virq=(16) \
030:   virq=(17) \
031:   pirq=+(9-11) pirq=+(5-6) \
032:   vdevs-size=32k \
033:   vtsc-offset \
034:   vcons-input \
035:   vmac=(0;0:0:0:0:0:4) \
036:   vdisk-wait=(3,1) \
037:   root=/dev/hda1 noirqdebug acpi=off console=ttyS0,9600 guest-initrd"
```

Figure 9.31: Excerpt of VLX's core configuration file

I/O ports and PCI devices. Now, the PCI configuration must be tuned to satisfy the initial requirements for the VMs. This is specified by the configuration statements at lines 9–13. Although it is not required as ⁎ is all inclusive, `vpci=(3,0,1)` and `vpci=(4,0,1)` are added. This is very useful for VM settings side-by-side comparison and for debugging. Line 8 (3,0,1) specifies the function 1 of PCI device 0 on bus 3. Similarly, line 9 (4,0,1) specifies function 1 of PCI device 0 on bus 4. But what are these devices? In fact, VLX is instructed that Linux_1 will control the GbE ports redirected to the MPCBL0040's front panel which is depicted in Figure 9.32. Figure 9.33 and Figure 9.34

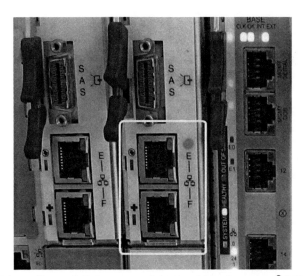

Figure 9.32: Redirected GeB ports of the Intel NetStructure® MPCBL0040

```
02:00.0 Ethernet controller: Intel Corporation 82571EB Gigabit Ethernet Controller
(rev 06)
02:00.1 Ethernet controller: Intel Corporation 82571EB Gigabit Ethernet Controller
(rev 06)
03:00.0 Ethernet controller: Intel Corporation 82571EB Gigabit Ethernet Controller
(rev 06)
03:00.1 Ethernet controller: Intel Corporation 82571EB Gigabit Ethernet Controller
(rev 06)
04:00.0 Ethernet controller: Intel Corporation 82571EB Gigabit Ethernet Controller
(rev 06)
04:00.1 Ethernet controller: Intel Corporation 82571EB Gigabit Ethernet Controller
(rev 06)
```

Figure 9.33: Filtering `/var/log/dmesg` **file to collect PCI mapping**

```
ACPI: PCI interrupt 0000:03:00.0[A] -> GSI 16 (level, low) -> IRQ 169 :eth0
ACPI: PCI interrupt 0000:03:00.1[B] -> GSI 17 (level, low) -> IRQ 193 :eth2
ACPI: PCI interrupt 0000:04:00.0[A] -> GSI 16 (level, low) -> IRQ 169 :eth1
ACPI: PCI interrupt 0000:04:00.1[B] -> GSI 17 (level, low) -> IRQ 193 :eth3
ACPI: PCI interrupt 0000:02:00.0[A] -> GSI 16 (level, low) -> IRQ 169 :eth4
ACPI: PCI interrupt 0000:02:00.1[B] -> GSI 17 (level, low) -> IRQ 193 :eth5
```

Figure 9.34: Filtering `/var/log/dmesg` **file to collect hardware interrupt mapping**

provide the GbE port's mapping to the PCI addresses. The SBC reference documentation can then be consulted to determine explicit details on the port mapping. To finalize the PCI setup of Linux_1, all the other GeB ports will be removed from control by Linux_1. This step is specified by lines 10–13, where a similar syntax is used as previously, with a leading exclamation point (!), which means that the device is disabled. At this stage, Linux_1 controls all the PCI devices, except the GeB ports A, B, C, and D. Implicitly, there are no PCI devices that are controlled by Linux_4 (default VLX behavior). Lines 27 and 28 specify adding GeB ports A and B.

In addition to assigning the PCI devices, the hardware interrupts that will be used by the VMs need to be specified. The virq entries of the configuration files control which hardware interrupts are visible to a VM. In line 6, `virq=(*)` grants Linux_1 access to all hardware interrupts (by default, no access is granted). Line 7, `virq=(17)`, makes the 17th hardware interrupt explicitly visible to Linux_1. A similar explicit declaration is employed as the PCI devices for debug and maintenance purposes. The reason why a `virq=!(x)` entry is not added, with x being the hardware interrupt line associated to GeB ports A and B, is due to the specific hardware configuration of the MPCBL0040 SBC. Indeed, the GeB ports associated with the PCI devices are sharing physical interrupts. In fact, all 0 PCI leaf functions are using IRQ line 16 and all 1 PCI leaf functions are using IRQ line 17. To accommodate this particular hardware architecture, Linux_1 and Linux_4 will share the same physical interrupts to control their GbE ports. Of course, this is specific to this SBC, and not to any sort of limitation in VLX. To conclude on the interrupts assignments, line 14 and the configuration line 31, `pirq=+(9-11) pirq=+(5-6)`, allows the VMM to use the PCI interrupt lines 5, 6, 9, 10, and 11 in addition to the lines that it has automatically enumerated. Note that IRQ 0 and 4 are reserved for the virtual timer and virtual UART. Regarding the timer, line 16, via the vtsc-offset option, instructs the VMM to compensate for delayed time tick delivery to a VM. This delay may occur when VMs are sharing a processor core, and accurate tracking of time is required. *Note*: VLX delivers all time ticks to the VMs.

Lines 18–20 specify the VBD mapping between the hardware block devices (the SAS drive's partitions) and the VBDs exposed to the VMs. In Linux_4, a front-end VBD driver relays the guest OS block device requests to the back-end driver. These requests are translated into transactions on the physical partition. The results are finally sent back to the front end. Mapping is done via the vdisk entries. The parameters between parentheses are split by a slash (/) into two groups. The left side group defines the destination VBD and the right side group defines the source physical block device. Consequently, in line 18, vdisk=(3,3,1/8,3,rw) instructs VLX that:

- This VBD will be available to VM number 3 (Linux_4 in our example),

- This VBD is /dev/hda1 (major number 3, minor number 1),

- This VBD is mapped to the physical partition /dev/sda1 (major number 8, minor number 1), and

- Linux_4 has read and write permissions to the mapped partition.

In line 36, with vdisk-wait=(3,1), VLX is instructed that Linux_4 has access to the virtual device, /dev/hda1. Note that this is also the name that the guest OS must use when it is booted in the VM. The -wait flag blocks the front-end VBD driver until the virtual disk becomes available in Linux_1. When the virtualized system is brought up, the boot process for Linux_4 requiring /dev/hda1 will be suspended until Linux_1 completes the initializations of /dev/sda1. Once this synchronization point is passed, Linux_1 and Linux_4 will execute in parallel.

Finally, the Virtual Ethernet Controllers (VEC) used by the VMs to communicate using a virtual Ethernet segment are defined. Lines 17 and 35 (vmac=(0;0:0:0:0:0:1) and vmac=(0;0:0:0:0:0:4)) specify that Linux_1 and Linux_4 are connected to the same virtual Ethernet segment (segment number 0 – the first parameter of the lists) and that Linux_1's VEC's Media Access Control (MAC) address is 1 and the MAC address of Linux_2's VEC is 2.

Now consider the remainder of the parameters that have not been discussed so far. Line 15 and 32, the vdevs-size entries fix the size of the shared memory space used for virtual device descriptors (32 KB in this case). The configuration command, vcons-input on line 34, simply specifies that Linux_4 will have the console input at boot time. The circulation of the console input between VMs is accomplished by using CTRL + SHIFT + Fn keys sequences sent via the terminal connected to the UART. The root entries (lines 21 and 37) provide boot options to the OS kernels. The significant options are:

- acpi=off – because no support for virtual ACPI interface is provided by VLX in version 1.0

- `console=ttyS0,9600` – matches the hardware and BIOS configuration for the COM port

- `guest-initrd` – specifies that a copy of the initial RAM disk image loaded by the GRUB boot loader is given to the guest system (`initrd` – it is a temporary root file system mounted during boot to support the two-state boot process).

9.4.2 Compute and Network I/O Performance

As mentioned earlier in this chapter, one can use virtualization to break through the non-scaling barrier of a certain class of distributed applications. To avoid any misunderstanding, it is important to be precise in stating that distributed telecom applications are scaling almost linearly with the number of computational nodes or blades composing telecommunication equipment. The issue is that these applications, at the compute node level, are usually not designed to take direct advantage of a multi-core multiprocessor system. Although it is always possible to introduce this support for such hybrid constructs (AMP with SMP nodes), such a change would require significant software modifications. Virtualization can be an elegant way to leverage multi-core processors and modern hardware, while running AMP legacy applications and legacy OSes. However, one can legitimately question the overall performance of such a virtualized system:

- Will a distributed application scale linearly with the number of VMs as it does with the number of physical blades?

- What will be the achievable I/O throughput?

I/O throughput is essential in telecom applications as it may theoretically suffer a considerable overhead, especially in the small packet sizes (64 B) typically used in these applications. With small packet sizes, the amount of time required to process the interrupts versus the amount of time needed to process the payload is significant. As always, where performance is concerned, nothing is better than measuring real application performance. Unfortunately, no well-selected industry standard benchmarks exist for measuring performance of such a virtualized system.

This last section presents performance measurements conducted using VirtualLogix™ VLX for the Network Infrastructure virtualization solution executing on an Intel

NetStructure® MPCBL0040 SBC. To answer the previous questions, two sets of measurements are conducted:

1. Evaluate the pure computational performance within a VM.

2. Evaluate the network I/O throughput of a VM.

In both cases, these performance numbers are compared to the performance numbers obtained without virtualization and using the same test applications. To demonstrate the validity of the use model to break through the SMP non-scaling barrier, the systems are configured in such a way that the non-virtualized performance represents an ideal SMP scaling configuration. Then, by aggregating the performance of each VM (each VM's performance being a proxy for a non-scaling, single threaded application) and comparing it to the ideal performance, the validity of our initial hypotheses can be confirmed.

SPEC CPU2000 V1.3 is employed to measure the compute performance. To be as representative as possible of telecom workloads (branchy, non-loop dominated code with almost no floating-point computations), only the integer portion (CINT2000) of the benchmark in rate mode is employed. In rate mode, it is possible to specify the number of users which translates into a number of instances of the benchmark that are executing in parallel. This configuration evaluates the compute throughput of a System Under Test (SUT). In addition, the use of SPEC CPU2000 minimizes I/O operations. In fact, only the processor and the memory subsystem – plus the chipset's memory controller hub – are stressed.

I/O throughput will be measured using a Layer 3 forwarding (IPv4) test with a SmartBits1 traffic generator and the SmartApplications* tester software (from Spirent® Communications* – http://www.spirentcom.com). Test parameters are set accordingly to the RFC 1242 and RFC 2544 standards. Simplified setup is depicted in Figure 9.35.

Performance measurements are obtained by executing and measuring the benchmark (CINT2000 in rate mode) score for two configurations – A and B. Configuration A is illustrated in Figure 9.36 and is composed solely of the MPCBL0040 SBC, running one instance of the Linux OS (RHEL4U3RC4-2.6.9-34) in SMP mode and therefore using four processor cores (two Dual-Core Intel® Xeon® LV 2.0 GHz processors). Configuration B is illustrated in Figure 9.37 and employs VLX to virtualize the MPCBL0040 SBC and manage four VMs. Each VM is running one instance of the same Linux OS (RHEL4U3RC4-2.6.9-34). The main difference is that each Linux is booted in single processor mode (UP). The same set of SPEC CPU2000 binaries is used with both configurations. It is important

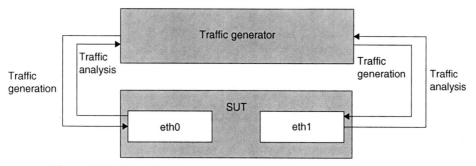

Figure 9.35: Simplified network I/O throughput measurement setup

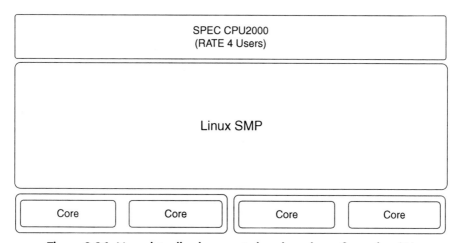

Figure 9.36: Non-virtualized compute benchmark configuration (A)

to note that the data shown in this section is not intended to illustrate the hardware and software peak performance, but is solely used for a relative comparison.

Based on the benchmark's profile and on knowledge of Intel® VT-x, the expectation is that the overall computational throughput of configuration A and B should be equal. This is essentially due to the fact that no particular events in the benchmark's workload are susceptible to initiating high rates of VM Exit transitions. Figure 9.38 summarizes the benchmark performance results. As expected, the sum of the CINT2000 rate mode scores of each VM is equal to the score of the non-virtualized SUT. The slight difference can be

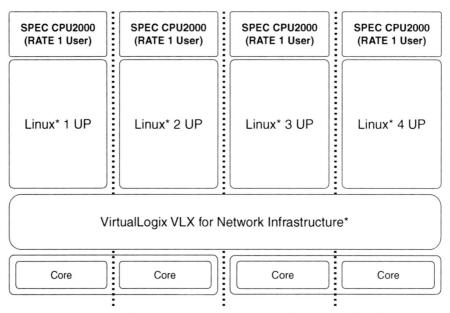

Figure 9.37: Virtualized compute benchmark configuration (B)

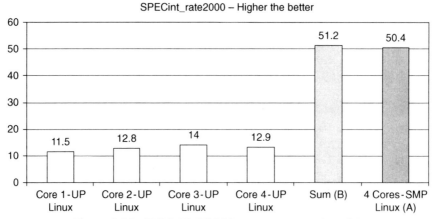

Figure 9.38: SPEC CINT2000 rate scores on A and B

first explained by the benchmark's metric variability. Second, it can be attributed to the fact that in each VM, the single-processor kernel performs better than the multiprocessor kernel, which has less synchronizations to manage and has a straightforward processor management scheme.

The conclusion is that there is no real surprise on the pure computation side; the use of virtualization is valid for the compute part of the problem.

Now consider I/O throughput. The network traffic or workload used is the SmartBits[*] 6000B traffic generator which is equipped with two copper GbE SmartCards (SC). Each SC is equipped with two copper Ethernet ports. The generator sends fully characterized and deterministic network traffic (described later) to one port of one network interface card (NIC) of the SUT. Next, the SUT then forwards the received packets to the other port of the SC. In the experiment, the network traffic is bi-directional, which means that the same port handles egress and ingress traffic flows simultaneously. Failure is satisfied if a single packet is dropped by the SUT. The traffic generator automatically searches by dichotomy the maximum throughput. This setup can be replicated as many times as the SUT has available pairs of ports. Telecom applications executing in the control plane, typically employ general purpose processors to process small packets because small packet processing is extremely demanding in term of processor resources (because of the time devoted to interrupt processing). Therefore, throughput with packet sizes ranging from 64 to 1518 bytes was measured. In order to be complete, the tests were performed using the Internet Protocol (IP) and User Datagram Protocol (UDP). In comparing the network I/O throughput between a virtualized and a non-virtualized environment, two configurations are defined, configuration A and configuration B.

Configuration A is solely composed of the MPCBL0040 SBC, executing one instance of the Linux OS (RHEL4U3RC4-2.6.9-34 and e1000-7.0.33 network driver) in SMP mode (4 processors) and with IP forwarding activated (`echo 1>/proc/sys/net/ipv4/ ip_forward`). Four network ports out of the six available on the MPCBL0040 are connected to the traffic generator (using 2 SCs). Note that because the communication backplane in the ATCA chassis is used, two ZX7120 72-port Gigabit Ethernet non-blocking switches by ZNYX Networks[*] are employed [10]. Figures 9.39 and 9.40, and Table 9.4 comprise the setup details. Configuration B employs VLX to virtualize the MPCBL0040 SBC and manage two VMs, each executing one instance of the same Linux OS (RHEL4U3RC4-2.6.9-34 and e1000-7.0.33 network driver) in UP mode (1 processor core) and with IP forwarding activated (`echo 1>/proc/sys/net/ipv4/ip_ forward`). Four network ports out of the six available of the MPCBL0040 are connected to the same SCs of the traffic generator. The difference is that each VM is managing two of them via native Linux drivers. The ports' PCI addresses and associated

Figure 9.39: MPCBL0040 SBC backplane 4 GbE ports

interrupts are assigned to the VMs by partitioning. Figures 9.41 and 9.42 illustrate the high-level setup.

The next step is to verify that the switches do not alter the I/O throughput when backplane ports are used instead of front panel ports. This is accomplished by using the traffic in closed loop configuration sending and receiving the traffic through the switches. Figure 9.43 clearly shows that the impact from the switches is negligible and a total throughput of 4 Gbps is achieved using IP and UDP protocols. Figures 9.44 and 9.45 summarize the throughput measurements (IP and UDP throughput curves are difficult to distinguish on the graphs).

Consider now the IP traffic (the UDP traffic analysis is similar and will not be detailed). Please note that no special optimizations, such as using processor affinity on interrupt servicing routines or kernel optimizations were done. Here again, rather than seeking pure performance, the goal is to compare the configurations.

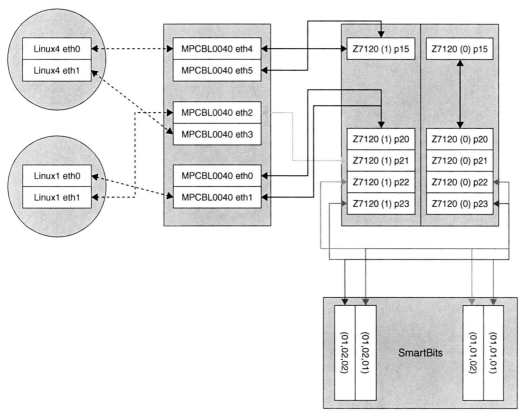

Figure 9.40: I/O throughput measurement setup details

Table 9.4: SUT I/O and traffic generator configuration

SmartCard (SC)	SC MAC	SC IP	Router's IP	VM	VM φ Iface PCI	VM φ Iface IRQ	VM Iface	VM Iface IRQ	VM Iface IP
(01,01,01)	00:00:00:00:00:01	1.1.1.2	1.1.1.1	Linux1	(3,0,0)	16	eth0	10	1.1.1.1
(01,01,02)	00:00:00:00:00:02	2.2.2.2	2.2.2.1	Linux1	(4,0,0)	16	eth1	10	2.2.2.1
(01,02,01)	00:00:00:00:00:03	3.3.3.3	3.3.3.1	Linux4	(2,0,1)	17	eth0	7	3.3.3.1
(01,02,01)	00:00:00:00:00:04	4.4.4.2	4.4.4.1	Linux4	(3,0,1)	17	eth1	7	4.4.4.1

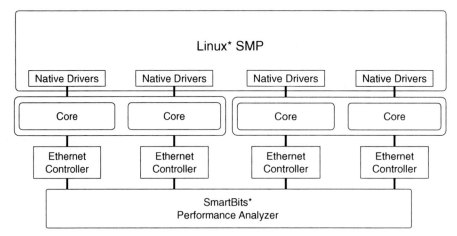

Figure 9.41: First I/O benchmark configuration (A)

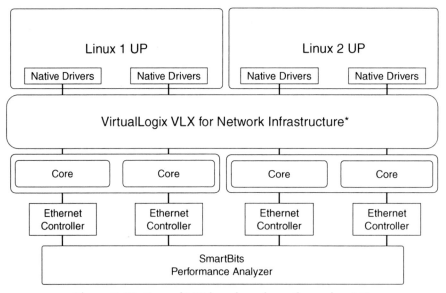

Figure 9.42: Second I/O benchmark configuration (A)

The throughput for Configuration A (labeled Linux SMP 4 NIC) represents how the non-virtualized SBC and the SMP kernel handles the test traffic. With the initial assumption that the VMM will introduce non-negligible I/O performance degradation, these curves should represent the targeted throughput that the aggregated throughputs of the VMs shall

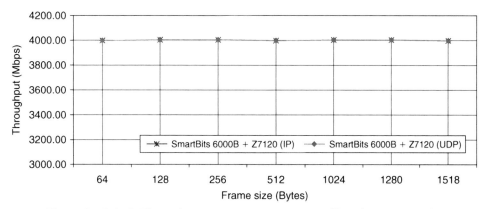

Figure 9.43: I/O Throughput measurement setup calibration/test results

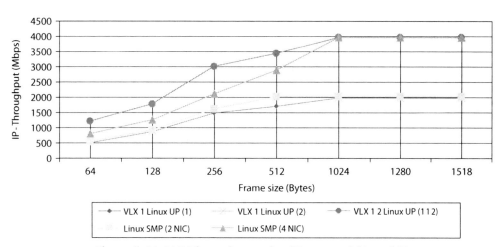

Figure 9.44: I/O Throughput using IP protocol (A and B)

match as much as possible. In addition to the 4 NICs configurations of A, the 2 NICs configuration to add a comparison point between a VM running a UP kernel and 2 NICs [labeled VLX+Linux UP (1) and VLX+Linux UP (2)] and an SMP kernel and 2 NICs (Linux SMP 2 NIC) are detailed.

The data first shows that the throughput of each VM is very similar (their throughput curves are difficult to distinguish on the graphs). Also it is apparent that each VM's

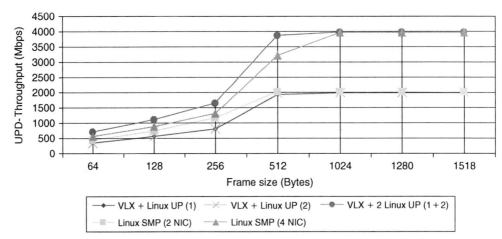

Figure 9.45: I/O Throughput using UDP protocol (A and B)

throughput curve is under the SMP 2 NICs curve, with the exception of the first data point where the VMs outperform A (see Tables 9.5 and 9.6 line *Delta 2 NICs*). This number meets expectation. A comparison between the aggregated throughput of the two VMs, using 4 NICs in total (labeled VLX+2 Linux UP (1+2)) and the SMP 4 NICs reference throughput of A reveals that the VMs are now outperforming A, especially in the smaller packets sizes. This is a phenomenon that was noticed earlier with the 64 B data point (see Table 9.5, line *Delta 4 NICs*). Note that on large packet sizes, throughput differences are minimal in 2 or 4 NICs configuration. This is important to mention as IT class applications, such as databases and mail servers, use larger packets than telecom applications. The performance deficit of the non-virtualized system versus the aggregated virtualized systems is essentially due to the better efficiency of the UP kernel to handle the network traffic. Yet again, keep in mind that no specific optimizations were attempted on either configuration. Nonetheless, the conclusion is still valid: the use of virtualization provides good I/O throughput performance.

Chapter Summary

This chapter provided a comprehensive view of virtualization and partitioning with application in embedded. A historical view was first provided followed by benefits of

Table 9.5: I/O Throughput using IP protocol (A and B)

Frame size (bytes)	64	128	256	512	1024	1280	1518
VLX+Linux UP (1)	600.00	886.23	1,500.00	1,710.61	1,986.67	1,986.25	1,987.07
VLX+Linux UP (2)	600.00	886.23	1,500.00	1,710.61	1,986.67	1,986.25	1,987.07
VLX+2 Linux UP (1+2)	1,200.00	1,772.45	3,000.00	3,421.23	3,973.37	3,972.51	3,974.14
Linux SMP (2 NIC)	531.65	902.44	1,604.65	2,000.00	2,000.00	2,000.00	1,999.99
Linux SMP (4 NIC)	800.00	1,251.59	2,123.08	2,899.19	3,973.35	3,972.51	3,974.14
Delta 2 NICs	*12.86%*	*−1.80%*	*−6.52%*	*−14.47%*	*−0.67%*	*−0.69%*	*−0.65%*
Delta 4 NICS	*33.33%*	*29.39%*	*29.23%*	*15.26%*	*0.00%*	*0.00%*	*0.00%*

Table 9.6: I/O Throughput using UDP protocol (A and B)

Frame size (bytes)	64	128	256	512	1024	1280	1518
VLX+Linux UP (1)	350.00	550.19	812.96	1,934.55	1,986.67	1,986.25	1,987.07
VLX+Linux UP (2)	350.00	550.19	812.96	1,934.55	1,986.67	1,986.25	1,987.07
VLX+2 Linux UP (1+2)	700.02	1,100.42	1,626.08	3,869.46	3,974.18	3,973.54	3,975.37
Linux SMP (2 NIC)	428.57	718.45	1,150.00	2,000.00	2,000.00	2,000.00	1,999.99
Linux SMP (4 NIC)	549.92	874.45	1,323.74	3,224.24	3,973.35	3,972.51	3,974.14
Delta 2 NICs	*−18.33%*	*−23.42%*	*−29.31%*	*−3.27%*	*−0.67%*	*−0.69%*	*−0.65%*
Delta 4 NICS	*21.44%*	*20.54%*	*18.59%*	*16.67%*	*0.02%*	*0.03%*	*0.03%*

virtualization. Several challenges with implementing virtualization on IA-32 architecture processors are discussed along with their solutions. Finally, a case study employing VirtualLogix™ VLX for the Network Infrastructure is detailed showing how virtualization of an embedded system is accomplished along with the performance benefits.

Related Reading

Additional reading material can be found on virtualization in the literature and on the Internet. "Applied Virtualization Technology" [11] by Campbell and Jeronimo surveys in-depth virtualization use models. On the technical side, "Virtual Machines: Versatile Platforms for Systems and Processes" [12] provides a broad view of VMs, including virtualization of course, but also addressing topics such as binary translation, emulation, etc. It is a reference text for those desiring a broad understanding of virtualization's multiple facets, including implementation techniques. As discussed earlier in the chapter, hardware support for processor virtualization is and will be extended in the future to continuously improve the performance and the security level of VMMs and VMs. The I/O is obviously the most critical one for the future. The "Intel® Virtualization Technology for Directed I/O" [13] technology article surveys established and emerging techniques for I/O virtualization. Its companion reference document is the "Intel® Virtualization Technology for Directed I/O" [14]. These documents are a good starting point to understand and prepare for the next major improvement in virtualization on Intel architecture. Similarly, assuming that you have a PCI-SIG membership, "Address Translation Services 1.0 Specification and Single Root I/O Virtualization 1.0 Specification" [15] will provide a thorough presentation of the PCI-SIG I/O Virtualization, an industry-wide initiative for I/O virtualization. Still among the technical reference publications, the "Intel Virtualization Technology Specification for the IA-32 Architecture" [16] is highly recommended. In this chapter, Explicit Parallel Instruction Computer (EPIC) architecture and its hardware support for processor virtualization, called VT-I, was not discussed. More can be found on VT-i in the "Intel Virtualization Technology Specification for the Intel Itanium Architecture" reference document at the same address. Readers interested in paravirtualization can review the reference article "Xen and the Art of Virtualization" [17] and review the material from Paravirtual [18].

References

[1]   Ersatz-11, http://www.dbit.com

[2]   REVIVER-11S, http://www.comwaretech.com/PDP-11/PDP-11-emulator.html

[3] Unified Extensible Firmware Interface, http://www.uefi.org/home

[4] S. King, P. M. Chen, Y. Wang, C. Verbowski, H. J. Wang and J. R. Lorch, *SubVirt: Implementing Malware with Virtual Machines*, http://www.eecs.umich.edu/virtual/papers/king06.pdf

[5] Advanced Telecom Computing Architecture, http://developer.intel.com/technology/atca/index.htm

[6] Service Availability Forum, http://www.saforum.org

[7] *Intel® 64 and IA-32 Architectures Software Developer's Manual Volume 3A: System Programming Guide*, Part 1, 154pp.

[8] PCI-SIG I/O Virtualization Specifications, http://www.pcisig.com/specifications/iov

[9] Intel NetStructure® MPCBL0040 SBC, http://developer.intel.com/design/telecom/products/cbp/atca/9825/overview.htm

[10] ZX71000 PICMG ATCA Hub Board, http://www.znyx.com/products/hardware/zx7100.htm

[11] S. Campbell and M. Jeronimo, *Applied Virtualization Technology*, Intel Press, US, 2006, ISBN-10: 0976483238.

[12] J. E. Smith and R. Nair, *Virtual Machines: Versatile Platforms for Systems and Processes*. San Francisco, CA: Morgan Kaufmann, 2005, ISBN-10: 1558609105.

[13] Intel® Virtualization Technology for Directed I/O, http://www.intel.com/technology/itj/2006/v10i3/2-io/1-abstract.htm

[14] Intel® Virtualization Technology for Directed I/O, http://download.intel.com/technology/computing/vptech/Intel(r)_VT_for_Direct_IO.pdf

[15] Address Translation Services 1.0 Specification and Single Root I/O Virtualization 1.0 Specification, http://www.pcisig.com/specifications/iov

[16] Intel Virtualization Technology Specification for the IA-32 Architecture, http://www.intel.com/technology/platform-technology/virtualization/index.htm

[17] P. Barham, B. Dragovic, K. Fraser, S. Hand, T. Harris, A. Ho, R. Neugebauer, I. Pratt and A. Warfield, Xen and the Art of Virtualization, SOSP'03, 2003, http://www.cl.cam.ac.uk/research/srg/netos/papers/2003-xensosp.pdf

[18] Paravirtual, www.paravirtual.com/default.htm

## CHAPTER 10

Getting Ready for Low Power Intel Architecture

Key Points

- The introduction of new Intel Architecture processors designed for low power enable a new class of Mobile Internet Devices (MIDs).

- Battery Life Toolkit (BLTK) helps characterize power utilization of different design decisions.

- Embedded debugging requires special-purpose equipment and features to aid phases of development, spanning hardware platform bringup, OS and device driver debugging, and application debugging.

Intel Architecture processors are suitable for a wide range of embedded applications spanning high performance server class products to low power mobile products. The low power capabilities of Intel Architecture processors targeting the notebook market segments reach down to devices requiring 7–10 W thermal design power (TDP); however, there are embedded market segments that require processors that consume even less power. New Low Power Intel Architecture (LPIA) processors are being designed to enable embedded developers to take advantage of general purpose and full-featured 32-bit and 64-bit processors in these market segments.

Embedded developers have much to gain by considering LPIA in their future designs including:

- Software ecosystem supporting Intel Architecture – a broad ecosystem provides a rich set of widely used software components from which to choose for deployment in your application.

- Compile once, run everywhere – development costs can be reduced by limiting the number of architecture and OS combinations targeted by your embedded applications.

Many embedded market segments can take advantage of LPIA processors. If a network device manufacturer employed multi-core x86 processors in their high-end networking products, LPIA could be used in their lower end, more power-constrained products and these two products could share software components. The automotive infotainment market segment can gain access to the rich x86 software ecosystem by employing LPIA in their thermally constrained environments. Mobile Internet Devices (MIDs) using LPIA offer Internet browsing based on the same browser software employed in the desktop/server market segments and can therefore offer a high level of compatibility with existing Web content on a mobile, long battery life device.

This chapter details what embedded developers need to know in order to prepare for LPIA processors. An explanation of the primary architectural change in LPIA processors is shared. This change is the reversion from out-of-order execution seen in every IA-32 architecture processor since the Intel® Pentium Pro processor to in-order execution. Second, a case study focusing on battery life evaluation is discussed. Battery Life Toolkit (BLTK) is employed to show the power benefits of a high performance compiler in an image processing application. Miscellaneous software techniques to reduce power utilization are shared. Finally, aspects of embedded kernel debugging are discussed.

Initial LPIA processors are single core so you may wonder why this topic is discussed here. Long term, it is very likely that LPIA processors will consist of multiple processor cores in one form or another, either in the form of a heterogeneous mixture of processor cores or the familiar homogenous multi-core processors of their higher power brethren.

Mobile Internet Devices – MIDs are a new category of embedded device that logically resides between high-end smartphones and laptop computers in terms of size and capabilities with regard to the Internet. One motivation for these devices is apparent from a comparison of MID with smartphone and notebook devices in Figure 10.1. MIDs possess many of the portability features of Smartphone devices while having more of the Internet viewing conveniences of a notebook system (larger screen and higher level of Internet compatibility).

Category	Smartphone	MID	Notebook
Screen Size	2″–3″ display	4″–6″ display	10″–17″ display
Boot time	Instant-on	Instant-on	Seconds to minutes
Internet compatibility	Moderate	High	Highest
User Interface	Simplified GUI	Simplified GUI	Full-featured GUI
Performance	Moderate	High	Highest
Battery life	Days	Days	Hours

Figure 10.1: MID comparison

10.1 Architecture

The current LPIA processors are the Intel processor A100 and A110, which are low power derivatives of the Intel® Pentium® M processor built on a 90 nm process and possessing a reduced 512 KB L2 cache.

Introduced in early 2008, the Intel® Atom™ processor is a new design built with a focus on lower power utilization and incorporating the following changes from recent IA processor designs:

- In-order processing
- Simplified decode and branch prediction
- Fewer specialized functional units

One of the critical changes is the employment of in-order execution instead of out-of-order execution. The other two changes reduce the power used by the processor core and go hand in hand with the change to in-order execution. In-order execution constrains the set of instructions that can be executed in parallel so a less sophisticated decode stage is sufficient to feed the execution stage. Likewise, fewer functional units are required to process the smaller set of instructions ready for execution at a given time. The processor is still superscalar just not to the degree of the Intel® Core™ microarchitecture.

10.1.1 In-order Execution

In-order and out-of-order execution describes how instructions are processed by the execute stage of the processor pipeline. In general, out-of-order execution results in

```
/* Sample C code */
a = b * 7;
c = d * 7;

/* Assembly language representation */
A   movl b, %eax
B   imull $7, %eax
C   movl %eax, a
D   movl d, %edx
E   imull $7, %edx
F   movl %edx, c
```
Figure 10.2: Execution order code sample

higher performance than in-order execution; however, this requires more circuitry to support. In a processor that supports in-order execution, the execute stage processes the instructions in the order specified in the executable. This order was either defined by the compiler during compilation or by the assembly language writer. In a processor that supports out-of-order execution, the execute stage can process instructions in a different order than was specified by the executable and can dynamically reorder instruction execution based upon resolution of dependencies. The following set of examples illustrates the differences between out-of-order and in-order execution. Consider the sample C code and equivalent assembly language representation for two multiply operations in Figure 10.2.

For easy reference a letter has been added before each assembly language instruction. Notice the dependencies between instructions A, B, and C and the dependencies between instructions D, E, and F. Instruction C cannot execute until the result of instruction B is available and instruction B cannot execute until the result of instruction A is available. The same relationship exists between instruction F, E, and D. There is no dependency between any instruction in stream 1 (instructions A, B, and C) and stream 2 (D, E, F). In the following figures, time is denoted in serially incrementing periods with the assumption that each instruction takes one time period to execute. In addition, to help simplify the explanation, it is assumed that no other processor-pipeline constraint has an effect. In a real processor pipeline, decode bandwidth limitations and execution unit limitations also impact when instructions can execute.

Time	1	2	3	4	5	6	7	8
Stream 1	A	B	C					
Stream 2	D	E	F					

Figure 10.3: Out-of-order execution

Time	1	2	3	4	5	6	7	8
Stream 1	A	B	C					
Stream 2			D	E	F			

Figure 10.4: In-order execution

Figure 10.3 shows how the two instruction streams would execute on a processor that supports out-of-order execution. The nice property of out-of-order execution is that the processor can execute an instruction when the instruction's dependencies are satisfied instead of being forced to execute the instructions in the explicitly laid down instruction order.

Figure 10.4 depicts the order of execution for the two streams on a superscalar processor that supports in-order execution. Notice that even though instruction D is not dependent on any instruction in Stream 1, instruction D does not execute until time period 3.

Unfortunately, this is the drawback to in-order execution and in general it can be said that in-order execution is less efficient than out-of-order execution as measured by the ratio of instructions retired divided by clock cycle (IPC). The performance impact of this inherent limitation can be reduced by effective compiler optimization in the form of an instruction scheduling phase that is keenly aware of the processor pipeline and schedules instructions for in-order execution.

```
/* Assembly language representation */
A  movl b, %eax
D  movl d, %edx
B  imull $7, %eax
E  imull $7, %edx
C  movl %eax, a
F  movl %edx, c
```

Figure 10.5: Instruction scheduling for in-order execution

Instead of relying upon out-of-order execution to dynamically choose instructions for execution based upon resolution of dependencies, a compiler can effectively schedule for these dependencies ahead of time. Figure 10.5 shows a new ordering of instructions that enables a processor that supports in-order execution to exhibit the same pipeline behavior as the processor that supports out-of-order execution depicted in Figure 10.3. The instructions from the two different streams are intertwined so that they can execute together and take advantage of the superscalar features of the processor. The instruction order specified in the executable is the same as the dynamic reordering determined by a processor that supports out-of-order execution. In this regard, the compiler and its instruction scheduling phase is even more important for processors that support in-order execution than for processors that support out-of-order execution.

The previous examples were simple; a compiler that optimizes for a real processor supporting in-order execution must perform these additional actions:

- Account for instruction latencies and throughputs,

- Model hardware dependencies between instructions,

- Model stages of pipeline, which may cause delays in execution, and

- Discover best possible ordering with respect to all above.

In summary, the architecture of the next generation LPIA processor supports in-order execution, which means instructions are executed in the same order as listed in the executable. There is a performance impact from in-order execution as compared to out-of-order execution; however, a compiler can lessen the impact by effective instruction scheduling. Finally, one thing to keep in mind: LPIA processors are backward compatible with other x86 processors so compilers that do not schedule for in-order execution can create executables that will run correctly but some performance is left on the table.

Case Study: Battery Life Toolkit

Battery Life Toolkit (BLTK) was briefly reviewed in Chapter 3. BLTK helps quantify the effects of different system design characteristics on battery life and power utilization that are primary concerns in mobile and thermally constrained embedded systems. Improvements in power utilization lead to several advantages including:

- Longer system usage between battery recharges, and

- More processing power within a given power and thermal envelope.

How does one measure the power savings afforded by a design decision? Standardized benchmarks that quantify system power performance are available; however, you may have a specific application to characterize. BLTK can be used in these cases.

This case study offers insight into how to effectively use BLTK to characterize power utilization of various design decisions. In this case, the design decision concerns which of several compiler and compiler optimizations to use in order to maximize power and performance.

Compiler Differentiation on Power

The performance impact of compilers and compiler optimization is understood; there are many examples in previous chapters that offer proof. Conversely, the impact on power utilization by using aggressive compiler optimization is not as clear. For example, envision a scenario where compiler optimization resulted in an application that executed faster, but employed more power-hungry instructions. The application may run faster, but overall power utilization may be higher. Another possible outcome is that compiler optimization would eliminate redundant instructions or operations and net a significant power savings. Without measuring the actual benefits of compiler optimization, there is no conclusive answer.

Two experiments are conducted employing BLTK to determine if aggressive compiler optimization benefits power utilization. Both experiments employ gcc 4.1.1 as the default compiler and the Intel C++ Compiler version 10.0 as the optimizing compiler. The test system is a notebook system with an Intel® Pentium® M processor executing at 1.6 GHz, 512 MB of memory, and executing a custom kernel based upon Fedora Core 6 and kernel version 2.6.18. The experiments use POV-Ray, a ray tracing application, as the benchmark application and two different compiler and compiler optimization settings to answer the following:

- How long does the battery last when rendering an image at a constant interval?

- How many renders can be completed before the battery needs to be recharged?

Application areas such as DVD playback motivate the first question. In the case of DVD playback, there is an appreciable limit on the number of frames per second to decode. After a certain level, decoding and rendering more frames per second does not provide visual improvement because the human eye has limits on noticing such differences. In addition, it does not make sense to decode more frames per second to speed up the playback of the movie. Instead, the benefit from a more highly optimized application is to use less system resources to deliver the required amount of rendering response and thereby increase the length of time before the battery drains. Experiment #1 attempts to answer this question.

Experiment #1 compares performance of the application compiled by the default compiler and optimization settings against the optimizing compiler with its set of aggressive optimization settings. BLTK is modified to execute the POV-Ray application at a regular interval, which is the average time the application compiled by the default compiler takes plus 3 s of idle time. The application compiled by the optimizing compiler takes less time; in this case BLTK still starts execution of the application at the same regular interval as the default compiled version. The system is allowed to idle longer as a result of the faster execution before the next render begins. With both versions, BLTK records the number of renders before the battery drains and the system halts.

The second question speaks to flat out performance: just how much work can be accomplished before the battery needs a recharge? Some applications have a fixed amount of work per unit that needs to be accomplished. A more optimized version of the application can result in completing more units of work with the same amount of power as long as the optimized version does not significantly increase the amount of power used per unit. Experiment #2 attempts to answer this question.

Experiment #2 compares the POV-Ray application compiled by the default compiler and build options against the application compiled by aggressive compiler optimization. The same compilers and options are used as in the previous experiment. In this experiment, both versions render the image and then idle for 3 s. Since the optimized version completes a given render faster, it should iterate and render the image more times than the default version before the battery drains.

Building POV-Ray

Using POV-Ray as the benchmark application requires it to be built from source. For these experiments POV-Ray version 3.1g is used. Download, build, and execute steps are detailed in Table 10.1. The makefile for POV-Ray required the removal of the -m386 option from the setting for CFLAGS. The default optimization settings for POV-Ray are -O6 -finline-functions -ffast-math.

Building the optimized version requires conducting a profile-guided optimization of the application. In addition, automatic vectorization and interprocedural optimizations are used. Table 10.2 summarizes the steps to build the optimized version of POV-Ray.

Table 10.1: POV-Ray default build steps

Download POV-Ray [1] and uncompress	`mkdir wl_povray`
	`cd wl_povray`
	`tar -zxvf povuni_s.tgz`
	`tar -zxvf povuni_d.tgz`

Build libpng	`cd source/libpng`
	`make -f ./scripts/makefile.lnx clean`
	`make -f ./scripts/makefile.lnx`

Build POV-Ray	`cd ../unix`
	`make clean`
	`make newxwin`

Execute POV-Ray with input data	`./x-povray`
	`-I../../scenes/objects/prism2.pov +d -x`
	`+v1 +ft +mb25 +a0.300 +j1.000 +r3 -q9`
	`-w1024 -H768 -S1 -E768 -k0.000 -mv2.0`
	`+b1000 +L../../include`

Table 10.2: POV-Ray optimized build steps

Set the compiler environment	`source <install path to icc>/bin/iccvars.sh`
Set dynamic library path	`export LD_LIBRARY_PATH=<path to wl_povray>/povray31/source/libpng:<inst all path to icc>/lib`

Build libpng	`cd <path to wl_povray>/povray31/source/libpng`
	
	`make -f ./scripts/makefile.lnx clean`
	
	`make -f ./scripts/makefile.lnx CC=icc`

Build POV-Ray for profile generation	`cd ../unix`
	
	`make clean`
	
	`make newxwin CC=icc CFLAGS="-I. -I..`
	`-I../libpng -I../zlib -I/usr/X11R6/include`
	`-DCPU=686 -c -prof-gen`

Execute POV-Ray with input data for profile generation	`./x-povray -I../../scenes/objects/prism2.pov`
	`+d -x +vl +ft +mb25 +a0.300 +j1.000 +r3`
	`-q9 -w1024 -H768 -S1 -E768 -k0.000 -mv2.0`
	`+b1000 +L../../include`

Build POV-Ray with profile data	`make clean`
	
	`make newxwin CC=icc CFLAGS="-I. -I..`
	`-I../libpng -I../zlib -I/usr/X11R6/include`
	`-DCPU=686 -c -prof-use -xB -ipo -O2"`
	`LFLAGS="-L../libpng -lpng -L../zlib -lz -`
	`lm -L/usr/X11R6/lib -lX11 -prof-use -xB -ipo`
	`-O2"`

BLTK Modifications

BLTK includes a number of sample workloads already integrated into the tool such as a Linux kernel build and MP3 player. Integrating POV-Ray requires a number of changes to BLTK including replacing one of the sample workloads. The modifications and steps necessary to execute POV-Ray using BLTK to benchmark for power utilization are summarized in Table 10.3.

Table 10.3: BLTK modifications and instructions

Download BLTK [2] and uncompress	`tar -zxvf bltk-1.0.7.tar.gz` `cd bltk`
Modify `wl_game/bltk_game.sh`	Remove commands to execute game and replace with command to execute POV-Ray. Added command: `--compiler=<compiler>` where <compiler> is either *default* or *optimized*
Modify `bltk/tools/bltk/main.c` Modify `bltk/include/bltk.h`	Remove option and code associated with game benchmark and replace with option to run default or optimized build of POV-Ray
Modify `bltk/tools/bltk/bltk_get_hdparm.sh`	`set HD_NAME=hda`
Build BLTK	`make` `make su`
Copy `wl_povray` into `bltk` directory structure	`cp -r <path to directory>/` `wl_povray bltk/`
Execute BLTK	`./bltk -compiler=default` Unplug AC cord when instructed and wait until battery drains

Obtain results	`cd bltk`
	`bltk_report <results dir>`
	`cd <results dir>`
	View files Report and Score

The modifications to `bltk/bltk_game.sh` included adding a command line option to specify which version of POV-Ray to execute. The version must be specified because Experiment #1 requires different idle times. The idle time between iterations of the benchmark compiled by the default compiler is set to 3 s. The idle time of the benchmark compiled by the optimized version is computed by subtracting the execution time of the optimized version from the default version and then adding 3 s. In my experiments, the difference in execution time for one render between the default version and the optimized version is 23 s; therefore, the idle time for the optimized version is set to 26 s.

```
Rating
Battery Rating : 79 min (01:19:00)

cycle
base  score
caliber
T:  N       time        work       delay     response    idle     score
S:  21      149.97      149.97      0         149.97      0        0
S:  22      147.24      147.24      0         147.24      0        0
S:  23      147.36      147.36      0         147.36      0        0
S:  24      147.36      147.36      0         147.36      0        0
S:  25      151.89      151.89      0         151.89      0        0
S:  26      147.73      147.73      0         147.73      0        0
S:  27      149.51      149.51      0         149.51      0        0
S:  28      151.41      151.41      0         151.41      0        0
```

Figure 10.6: BLTK output

BLTK Results

The key output from an execution of BLTK is the battery results and the number of renders completed, which are reported in the files Report and Score. The key excerpts from the files appear in Figure 10.6. The Battery Rating entry indicates how long the battery lasted before being drained. The entry labeled "N" (from the file Score) indicates the number of renders completed during this time. The value of the "N" entry that appears at the bottom of the file indicates the total number of renders complete. The sample output in Figure 10.6 indicates the battery lasted 79 min before draining and rendered the image 28 times.

The results for Experiment #1 and Experiment #2 are summarized in Figure 10.7. The result for a given version is the average of 10 executions of BLTK.

The results on the left side (Experiment #1) indicate that the optimized version extended the battery life by 9% as a result of the more efficient execution, which allowed the system to idle for longer periods of time. The results on the right side (Experiment #2) indicate that the optimized version completed 18% more work before the battery drained.

The experiments and results shared are specific to compiler optimization and results on a notebook computer; however, the expectation is that BLTK can be used to test the effects of many design decisions on battery life for systems employing LPIA processors.

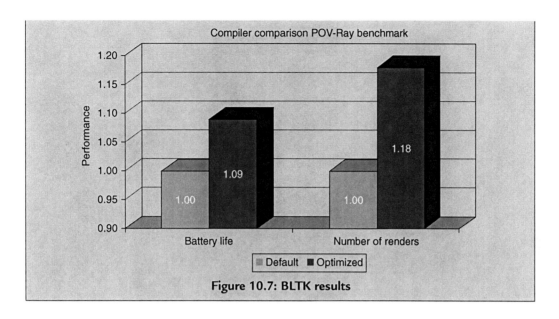

Figure 10.7: BLTK results

Miscellaneous Techniques for Energy Efficient Software

This section briefly discusses techniques on the following topic areas to improve battery life in mobile devices:

- Low-versus-high-granularity timer

- Hard drive block reads

- Compressed data (WLAN)

- LCD panel brightness

Again, these results were obtained on laptop systems but are general enough to apply to LPIA processor-based systems when they become available.

Recent Embedded Linux kernels offer higher resolution timers that can be used, for example, to collect application run-time statistics; however, use of these timers in power sensitive devices has a cost. Figure 10.8 shows the platform and processor power costs for a sample application employing timers of different granularity. On the sample application, employing a timer with a resolution of 100 ms saved about 1 W of platform power compared to timers of 1 and 10 ms granularity [3].

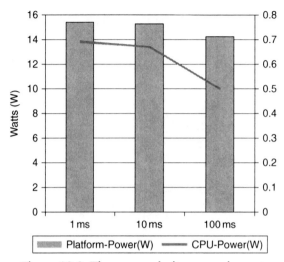

Figure 10.8: Timer granularity power impact

Block Size	CPU_ENERGY (mWH)	DISK_ENERGY (mWH)	PLATFORM_ENERGY (mWH)
1b	9658.77	1056.0	21,738.03
8b	1336.18	192.32	2712.85
1 KB	24.93	13.76	106.97
4 KB	24.05	13.56	104.92
8 KB	23.27	13.23	99.99
16 KB	22.46	12.98	98.05
32 KB	22.49	12.85	98.04
64 KB	22.50	12.96	98.05

Figure 10.9: Hard drive power utilization

Power utilization of a hard drive is affected by the average size of block reads. In general, power savings result from reading larger sized blocks and subsequently buffering them to memory for use afterward. Figure 10.9 details CPU, disk, and platform energy used for reading a file with a size of 1 GB in different block sizes from 1 byte to 64 KB. Minimize the number of explicit reads from disk by reading and buffering large regions in memory where possible [4].

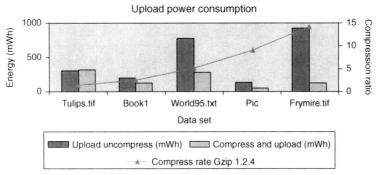

Figure 10.10: Upload power comparison

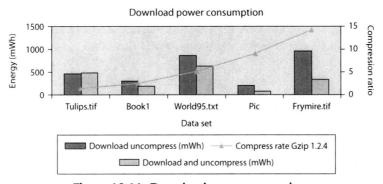

Figure 10.11: Download power comparison

A natural power trade-off in downloading and uploading data to and from a wireless LAN is the amount of compression used on the data. Does the act of compressing or decompressing a file use more power than waiting a bit longer while an uncompressed file is either uploaded or downloaded? Figure 10.10 shows a power consumption comparison for uploading a number of files between a compressed version of the file and an uncompressed version of the file. The power measurement for the compressed version of the files also includes the power used during compression. The graph also indicates the compression level used for the compressed file. Figure 10.11 shows a power consumption comparison for the same files and compression levels, but for downloading and subsequently uncompressing the compressed version of the files. As you can see, for the files that were compressed more than two times, significant power savings resulted from compressing the file before uploading and decompressing the file after downloading.

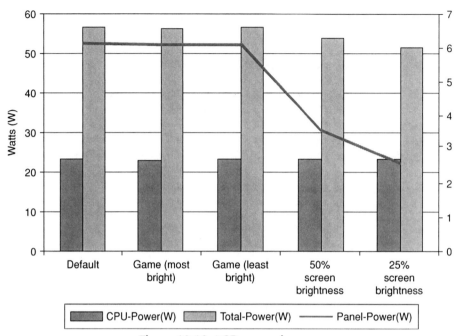

Figure 10.12: LCD power impact

In general, power utilization can be increased by transmitting files that have been compressed more than two times [5].

LCD panels consume a fair portion of platform power. Controlling the brightness of the screen is a technique to reduce platform power. Figure 10.12 shows the CPU, panel, and total power for a sample workload under a number of different scenarios. Current APIs to set brightness for games using Windows OSes have no impact on power utilization as evidenced by the similar results for a game measured at its brightest versus dimmest. Experimentation using the registry to programmatically set brightness to 50% and 25% result in savings of approximately 3 and 5 W respectively. Take advantage of software techniques to dim your embedded system's LCD display when idle [6].

10.2 Debugging Embedded Systems

Developing for embedded systems provides some unique challenges because in many cases you are developing across platforms; building your application on a host

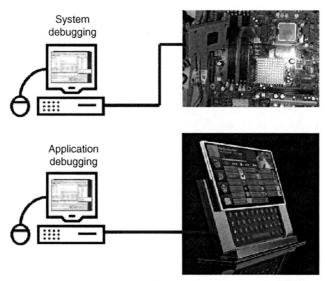

Figure 10.13: System debugging (JTAG) versus application debugging (TCP/IP)

environment that is different than the target environment where it is actually deployed and executed. It may even be the case that the target environment employs a processor architecture different than the one used by your host platform. Additionally complicating development, the target platform may be evolving as new features and functionality are added over the course of the development project. In other words, the target platform may not be entirely stable while you are performing your development.

Embedded developers therefore require tools to help debug their systems and require special functionality to overcome the inherent challenges. Figure 10.13 provides an example of two types of debugging, system and application, which will be detailed in the following sections. These sections include details on the following:

- History and trends in embedded systems debugging

- Overview of hardware platform bringup, OS, and device driver debugging

- Step-by-step instructions for OS kernel, services, and driver debugging

- Overview of application debugging

- Considerations for debugging multi-core processors

10.2.1 Brief History of Embedded Systems Debugging

Traditionally, the most common approach for early embedded cross-platform development was emulation via *in-circuit emulators* (ICE). A target processor would be mounted on a custom socket on top of which the ICE connector would be attached. The ICE actually contained a processor of its own and would be able to shadow most of the operations of the target platform. The main disadvantages of the approach were the enormous costs of these devices ($10,000+) and the need to build specialized platforms just for debugging purposes. Cost, of course, is a sensitive factor in embedded systems due to lower per-unit return compared to standard desktop/server systems.

In those days, the low cost alternative to ICE was the remote debug handler approach, which used a small bare-metal debug handler executed from a floppy disk or flashed as a hex file or plain binary onto the target BIOS boot flash. If your embedded system had tight code size and real-time requirements, you may have even integrated the debug handler into your RTOS image on the target platform. The problem with this approach is that the debug handler requires a dedicated communication stack and memory region on the target system which in effect steals these resources from the target application. A *debug handler* is an interrupt handler routine that stays idle until a debug event is detected on the target. As a result, the debug handler does incur system overhead and creates a timing impact which can cause undesired side-effects on systems that, for example, employ watchdog timers or other hard system timers to fulfill real-time or security requirements.

Today the debug handler approach is mostly used for embedded application development only. Many off-the-shelf embedded OSes (Palm[*] OS, Microsoft Windows[*] CE, HardHat Linux[*], etc.) have a built-in debug handler that enables remote debugging using TCP/IP or Universal Serial Bus (USB) protocols.

10.2.2 JTAG and Future Trends in Embedded Debugging

The most common approach used today in cross-debugging for hardware platform validation, bare-metal development, or embedded OS and device-driver adaptation employs *Joint Test Action Group (JTAG) probing*. JTAG is the common term used for the IEEE 1149.1 standard, titled "Standard Test Access Port and Boundary-Scan Architecture for test access ports used for testing printed circuit boards." JTAG was originally an industry group formed in 1985 to develop a method to test populated circuit boards after manufacture. The devised methods evolved into an IEEE standard in 1990. The first processor to support the JTAG standard was the Intel486™ processor. Since then, this standard has been adopted by electronics companies across the industry.

Initially designed for printed circuit boards, JTAG is now primarily used for accessing sub-blocks of integrated circuits and as a mechanism for debugging embedded systems. When used as a debugging tool, it enables a developer to access an on-chip debug module, which is integrated into the CPU via the JTAG interface. The debug module enables the programmer to debug the software of an embedded system. Figure 10.14 is a picture of such a target probe (top left corner) connected via JTAG to a motherboard.

JTAG revolutionized low level platform debug capabilities for the embedded software development community providing a direct serial link from the target processor to the

Figure 10.14: JTAG device and connector

host development machine. Products employing the JTAG interface for debugging exist today in two forms. First, there are the more advanced JTAG devices that frequently contain a MIPS* or ARM* architecture processor and memory of their own. These devices may provide additional features like extended execution trace or data trace. Usually, these devices offer remote debug connectability via TCP/IP. Second, the simpler JTAG debug interfaces usually contain a small application-specific integrated circuit (ASIC) that handles the communication protocol and translates the serial JTAG signals for transmission over a parallel (RS232) or USB connection.

With the advent of more complex SOC-embedded platform designs, the need for advanced debugging tools is increasing. SOC-embedded platforms will increasingly feature several dedicated and specialized heterogeneous processor cores allowing for highly parallelized application execution and short input reaction time paired with the efficiency of specialized communication and video processors already present in embedded systems today. Debugging these types of systems will be no easy task.

10.2.3 Hardware Platform Bringup

One of the main differences between development in the desktop/server market segments and development for an embedded system is embedded's high level of platform feature and form factor customization. For example, this customization may require developers to develop firmware and BIOS-type boot code to meet aggressive code and data size constraints. Another motivation for developing firmware and boot code is in heterogeneous multi-core/multiprocessor SOC platforms where one processor serves as the primary boot processor and the other processors require initialization. In both cases, the firmware and boot code require development and debug typically at a phase of the project where the embedded system itself is still being developed. When debugging a new platform possibly based on a pre-release processor and trying to figure out the best board layout and possible interactions between different platform components it is very important to have as much information as possible about the underlying platform at your fingertips.

A JTAG-based debugger is an ideal tool during the *hardware platform bringup* phase because it does not require the entire platform being fully functional and providing the services a piece of software or firmware executing on the platform would require. There is still run-time impact whenever you take control of the system remotely; however, this is a small trade-off for the capabilities provided.

The capabilities of a debugger that will be most valuable in this phase are:

- Full hardware and device register access,

- Basic assembly and source level debug capabilities for comparatively small statically linked objects (binaries, Intel hexfiles, Motorola s-record files), and

- The ability to remotely download these objects into target random access memory (RAM) or even write them into boot-flash or read-only memory (ROM).

One method of providing low level knowledge of execution is to give the developer full visibility of all registers, on-chip as well as memory-mapped and platform device registers. Figure 10.15 depicts one method of implementation with a bitwise graphical depiction of each of these registers and a short description of the register's purpose and the register configuration options. One advantage of this capability is in having the exact register layout at your fingertips and in most cases not even needing additional documentation beyond what the debugger describes.

Today typical embedded compilers generate ELF of COFF object files and allow for the generation of Dwarf-2/Dwarf-3 or Microsoft CodeView *debug symbol information.* Some legacy tools may still produce AOUT or OMF86 objects and may still require that a debugger be able to understand and translate STABS symbol information, but the number of these development environments is quickly diminishing. Outside of the Microsoft Windows targeted compilers, ELF Dwarf-2 objects and ELF Dwarf-2 is the *de facto* industry standard debug format. For further information on object formats and symbol information, the reader is referred to John R. Levine's *Linkers and Loaders* [7].

Once compiler-generated objects have been statically linked with or without basic boot code, you will then need to convert them into a format that can be readily understood by flash writers and can be downloaded and executed on your embedded target platform. Usually, this conversion from ELF to hex or binary formats is accomplished using a utility like GNU* objcopy.

If you are simply performing testing or validation in dedicated RAM, a great debugger feature is the ability to extract the symbol file from the aforementioned ELF file and then directly download the associated hex file or binary into target memory. Ideally, it would be just as simple with your boot ROM or flash. Some debuggers offer flash

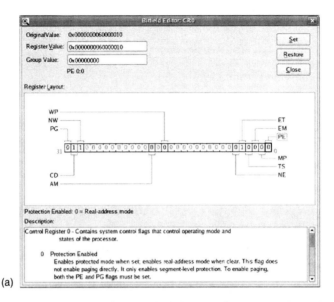

(a)

(b)

Figure 10.15: Bitfield Editor

support as part of the scripting language in the debugger or even allow specification via a configuration file of the memory layout of your platform designating which address ranges need to be protected, which are RAM, ROM, NOR flash, and NAND flash, or handled by filesystem support that may be part of the boot code.

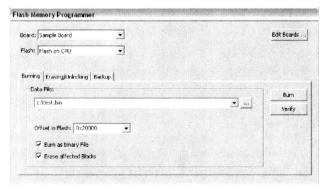

Figure 10.16: Flash Memory Programmer

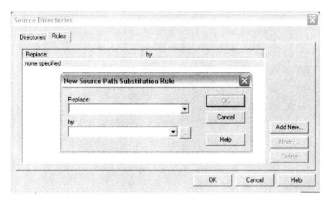

Figure 10.17: Remapping source directory path for symbol information matching

Alternatively, the debugger may also feature a fully functional flash writer integrated into the debugger GUI. Depicted in Figure 10.16, this approach maintains the clean separation between ROM and RAM and with its clear and user-friendly interface is similar in functionality to what is offered in a hardware-based flash writer device.

This approach makes flash writing independent of detailed debugger scripting knowledge, usable for developers without a great deal of experience in bare-metal software development, and due to ease of use is appealing to hardware validation engineers. Now that the image has been loaded into target memory it is just a matter of loading the associated debug symbol information into the debugger and associating the sources correctly.

Loading the debug symbol information is relatively straightforward because the information is contained in the linked ELF object; however, there are exceptions that are important to know. One exception is due to cross-development. The source path information stored in the debug symbol information may not match the actual source directory tree as seen from the debug host. The debugger therefore needs to offer the ability to remap the source tree. Figure 10.17 depicts a graphical interface for source code remapping available in typical embedded debuggers.

Now that you have your bare-metal code built, converted, downloaded into memory, and the debug symbol information in the debugger mapped, you are finally ready to debug your OS-free embedded test code … most likely using a JTAG connection.

10.2.4 OS and Device Driver Debugging

Once the embedded system can execute basic boot code or firmware, the next step is typically to determine if the OS can load and that all of the devices and I/O interfaces are recognized and operating correctly. Today's embedded systems are becoming richer in features and capabilities while at the same time becoming more and more complex. Hand in hand with this trend is a migration from proprietary small footprint RTOSes to the use of slightly modified full-featured OSes. As an example, the LPIA processors targeting application as MIDs employ what is essentially a desktop/server Linux OS with a modified graphical user interface and special device drivers to meet the specific needs of the device.

10.2.4.1 Case Study: OS Kernel Debugging

Consider an MID executing a kernel based upon the Mobile and Internet Linux Project [8]. This section employs the Intel® Debugger debugging on a LPIA processor; however, many of the detailed tips and techniques are generally applicable. In this case study, assume that you are connected to your embedded target device using a JTAG connector. Upon connection, the JTAG device takes control of the target and if you try to move the mouse on your target device or enter commands in a command shell it should not react – independent of whether you stopped directly after reset or whether your OS is already fully booted.

Most processors offer a choice between a reset JTAG connection and a hot-debug JTAG connection. In both cases the debugger will halt the target upon connection. The only difference is that in the first case the JTAG device will also issue a reset and stop on the reset vector after connection. In the second case the JTAG device will stop at whatever memory address the platform is currently executing. A reset JTAG connection is usually more desirable for platform bringup whereas a hot-debug connection is usually more desirable for device driver and OS debugging.

In either case, as soon as connection is established it is time to load the debug symbol information for the Linux kernel. After navigating to the vmlinux kernel ELF/Dwarf-2 file in the Linux kernel source tree that matches the Linux image running on the MID target, you would load the symbol information into the debugger.

In Intel Debugger script language format the command is:

```
LOAD /SEGMENT /DEBUG /NOLOAD /GLOBAL OF "/usr/linux-2.6.22.
i686/vmlinux"
```

To map the paths for the location of the sources associated with the debug symbol information on the debug host system correctly, it is necessary to tell the debugger where the top level directory of the source tree is located on the debug host system. Figure 10.17 shows one method of remapping. The equivalent script command is:

```
SET DIRECTORY /SUBSTITUTE=""""/usr/linux-2.6.22.i686/"
```

The Linux OS entry point where the OS kernel is unpacked and the actual OS boot process begins is usually called `start_kernel()`. To debug the OS boot process, you would thus set a hardware breakpoint at this function. This breakpoint should be a hardware breakpoint because the RAM memory is not fully configured this early in the boot process and you may run into problems using software breakpoints. To be on the safe side when it comes to single stepping through the boot code it is recommended to tell the debugger to treat all breakpoints as hardware breakpoints.

At this point, the target will stop execution after processor release and after running through the BIOS and the OS bootloader at the function entry point for `start_kernel()`. The source window for the debugger will open upon stopping at the breakpoint providing visibility into the Linux OS kernel sources. You should now be able to do almost everything that you are used to from "traditional" desktop application debugging, such as monitor variables, memory, module lists, call stacks, templates, prototypes, step through code, use complex breakpoints, and so on. By the time you reach `mwait_idle()`, the main schedule idle loop of the OS, the target is brought up all the way.

Figure 10.18: Page translation table entries

Some additional features that can be especially useful to identify memory leaks and stack overflows or execution flow issues include:

- A detailed breakout of the processor descriptor tables,

- Easily accessible GUI-supported page translation, and

- Readout with disassembly and source mapping of processor on-chip execution trace to verify where the actual code execution may have gone wrong.

If you are concerned about problems in your OS memory configuration you could either take a closer look at the descriptor tables or you could review a graphical representation of the physical-to-virtual memory translation by taking a look at the page translation table entries such as one depicted in Figure 10.18.

To determine if your memory mapping actually corresponds properly to the underlying physical memory, a direct mapping of the actual memory translation may suffice; an example of this translation is depicted in Figure 10.19. This detailed graphical representation shows how the virtual memory address maps to the underlying physical memory and whether at a specific memory address there may be a gap in the mapping or an allocation conflict of some kind.

A typical first step in isolating run-time problems in a system debug environment is to capture the failure condition, interrupt or exception that was triggered by the run-time problem. Processor execution trace can help determine the actual execution flow and where execution starts to deviate from the expected behavior. Figure 10.20 is a screenshot showing the execution trace feature of the Intel® Debugger.

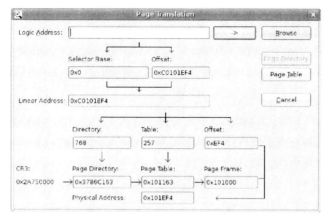

Figure 10.19: Page translation table

Figure 10.20: On-chip execution trace

Once you have pinpointed the code path and memory location associated with the issue, variable and register contents at the location can be analyzed to develop an understanding of where exactly in your code, some library function, the underlying memory, or caching model that is the cause. Execution trace is a very powerful tool that many embedded processors offer. In most architectures, readout of the trace memory does require going through the JTAG debug unit, again stressing the importance and value of JTAG device-based debugging.

10.2.4.2 Case Study: Debugging Services and Drivers

After OS kernel issues have been debugged satisfactorily, the focus shifts to periphery and device support. On Linux, support is typically implemented in the form of kernel modules that are either loaded by the OS during the boot phase or loaded at run-time by an application.

One challenge with debugging kernel modules for Linux kernel 2.6.xx and newer releases is that by default, the OS does not globally export information about the load location, initialization method, and cleanup method location in memory. In other words, the OS does not automatically provide the necessary information to determine how the debug symbol information needs to be remapped relative to the address space of the kernel module execution on the target device. This normally makes it very cumbersome to debug dynamically loaded kernel modules through a JTAG device or through a debug handler running on top of the target OS. The most straightforward solution to this issue is to implement a small kernel patch that exports the kernel module load information. This requires the OS kernel to be rebuilt to enable full kernel module debugging; however, this is a one time investment and the patch will be reasonably small with almost no memory and run-time overhead. Alternatively, a kernel module loaded on top of the running kernel can handle exporting the module load location information. Today the standard solution is to instrument each and every kernel module that needs to be debugged and remove the instrumentation before production. In my opinion, this is not a practical approach. Having the load location information handled in one central spot, whether it is the more user-friendly kernel module or the easier to implement kernel patch is preferable, leaving the developer more flexibility to debug a device driver should the need arise without too much additional debug ability implementation threshold.

Assuming the debug solution of choice allows building the kernel module with Dwarf-2 symbol information enabled, you could then instruct the debugger to load the symbol information as soon as the module load, module initialization, or module cleanup occurs. A screenshot of this step is depicted in Figure 10.21.

Once the debugger has taken control and halted the target, the driver of interest is loaded by the OS. Debug symbol information is loaded and you can pick the source file and function entry point of the portion of the device driver that you would like to debug. After setting the breakpoint and releasing control, the LPIA processor waits until a target event is hit causing the breakpoint to be reached and execution halted. Consider an example employing the dummy kernel module `scull.ko`.

In the debugger, instruct the Linux OS awareness plugin where to find the kernel module sources using the debugger console window and entering the following commands:

```
set directory "/home/lab/scull/\"

OS "setdir \"/home/lab/scull/\""
```

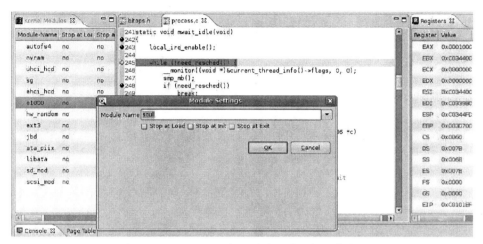

Figure 10.21: Kernel module selection dialog

Next, open the modules window and click the right button, and select "add" from the pull-down menu. You will then notice all of the other kernel modules already loaded during the listed boot process. In the next dialog box that appears, type in "scull" and select "Stop at Init." Now, release the target with the "run" command or button. Use PuTTy/SSH or Telnet to log onto the released and running target. After changing to the scull directory initialize the kernel module by typing:

```
./scull.init start
```

As expected, the debugger stops at the scull init method. Set a breakpoint at the `scull_read()` function and release the target once more. Next, send an echo to the /dev/scull0 device. The debugger halts and you can step through the function. Figure 10.22 displays a screenshot of the debugger halted in the scull kernel module.

To debug whether the kernel module is cleanly removed from memory after the device is deactivated you could also have decided to have the debugger stop and take control at the module cleanup method. For this example, the command to enter:

```
./scull.init stop
```

would cause the debugger to stop the target and upon halting, the program counter is set to the cleanup method of your scull kernel module. From there, debug of the actual unloading and disabling of your device can be accomplished.

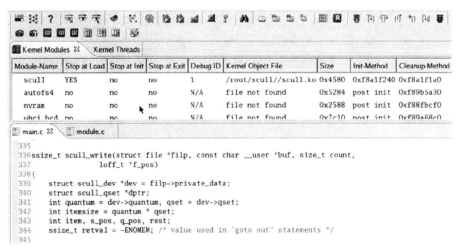

Figure 10.22: Debugging kernel module `scull.ko`

10.2.5 Application Debugging

The objectives and challenges of developing and debugging applications on an embedded platform are in many ways very similar to other platforms. The goal is to identify and fix program crashes and freezes as well as unexpected or unintended program behavior. An application debugger enables you to analyze all parts of your code and investigate program coding errors. With regard to debugging an embedded system, the previously discussed issue of cross-development impacts debugging by requiring use of a cross-debugger and execution of a remote server on the target.

One option before debugging on your target is to test the application on your development machine using a loopback connection to your host system. This may be particularly compelling if the board bringup and OS adaptation phases of development are not complete yet and thus the target platform is not fully available for your development efforts. This type of consideration again highlights the strong interdependency between software development and partly concurrent hardware development when dealing with embedded platforms.

As usual, for the initial debugging of an application it is recommended that you compile with debug information and no optimization. Only in later stages of application development should you work with optimized code. After building, the next step is to copy the application executable over to the target system for testing. Another option is to establish a shared drive between the host and target.

```
Intel(R) Debugger Remote Server for IA32 Linux, Version 0.6.6
Copyright (C) 2006-2007 Intel Corporation. All rights reserved.

Waiting for connection with "tcpip:2000"...
```

Figure 10.23: Debugger remote server waiting for connection

Figure 10.24: Debugger console

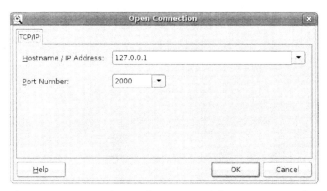

Figure 10.25: Configure the remote debugger connection

Initiating the debug session is accomplished by starting the debugger remote server (also called debug handler) on the target system. After launching, the debug handler typically identifies which TCP/IP port it is monitoring for the incoming debug request. Figure 10.23 shows the Intel® Debugger Remote Server output and identifies port 2000 for receiving debug session requests.

At this point, the application debugger on the host system can be started. The Intel Debugger console window is depicted in Figure 10.24.

Connecting the debugger to the remote server is accomplished by instructing the debugger which TCP/IP address and port the remote server is monitoring. This action is depicted in Figure 10.25.

Figure 10.26: Loading target application for debug

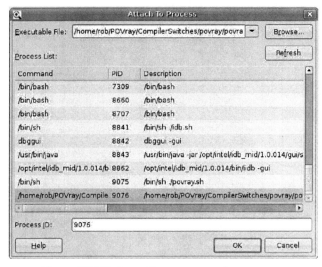

Figure 10.27: Attaching to a running process

The remote server will indicate on the target if a connection has been established and you can then proceed loading the application for debugging, an example of which is illustrated in Figure 10.26.

Now, you are ready to start your application debug session.

Another method of establishing a debug session is to connect to an already running process using the "Attach To Process" menu in the debugger as depicted in Figure 10.27.

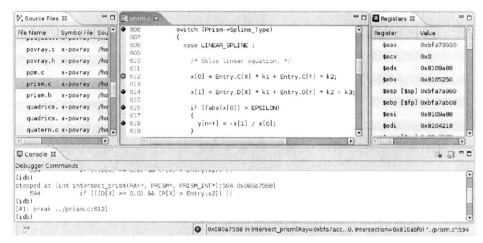

Figure 10.28: Application debug session

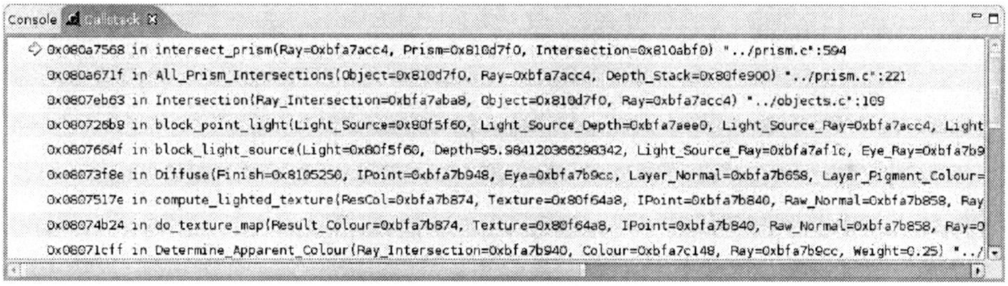

Figure 10.29: Callstack

The behavior of the debugger from now on will be very much like that of any standard debugger on a desktop/server computer with the ability to set breakpoints and step through the source code of the application. An example screenshot appears in Figure 10.28.

Similarly, visibility of your application callstack is available as depicted in Figure 10.29.

A typical debug session will therefore probably look something similar to the Intel Debugger GUI depicted in Figure 10.30.

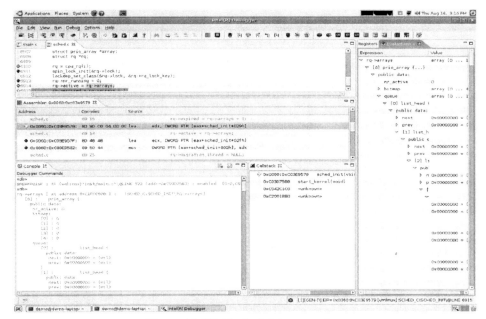

Figure 10.30: Typical debug session

This should look very similar to a standard debug window on desktop personal computers. Nevertheless, extended hardware awareness can be especially useful in embedded environments allowing:

- Visibility of the current values of the segment registers, general purpose registers, floating point registers, and the XMM registers all from one easily accessible list, and

- An application thread window which is particularly useful if your application is multi-threaded. It shows the thread currently under debug and information on the other threads that comprise application execution.

10.2.6 Considerations for Multi-core Debugging

With embedded systems becoming more and more complex and SOC configurations becoming more and more common, new challenges are presented when debugging these systems. For example, the following questions come to mind:

- How do you synchronize the execution flow over multiple processors and cores?

- How do you stop one thread on one core and how do you make sure a thread on another core can be stopped at exactly the same time if so desired?

- How do you synchronize breakpoints across multiple cores?

Adding to the complexity of debugging are the different types of cores that may reside in the system, such as a dedicated graphics core, a dedicated communications core, a dedicated vector machine, or dedicated network micro-engines. Frequently, on these types of systems a primary core (usually the application processor core) handles initialization of the other secondary cores. If the memory is shared between different processor cores, the debugger may be required to know the type of code stored so that it can properly interpret it for display.

Finally, on these multi-core systems you may have a unique secondary processor or platform specific set of templates or function API that is used to synchronize the cores and pass messages between the cores. For true multi-core debugging, the debugger should be able to be triggered on these synchronization calls and track the dependencies between the cores.

Looking at multi-core programming and development, we see standards developing and becoming more established for handling the synchronization challenges of involved (Microsoft* Threading Model, OpenMP*, and POSIX* Threads to name a few). Similarly, when handling multi-core debug environments there is no one industry standard technique to help debuggers synchronize between multiple cores in homogeneous and heterogeneous environments. This is a rapidly evolving field and many new and exciting developments are on the horizon.

10.2.6.1 JTAG

JTAG device system debugging is primarily used for board bringup and for debugging firmware that handles the communication between the primary application cores and specialized secondary cores. Implicit in this activity is the need to synchronize execution and breakpoints between the cores in the debugger without requiring the presence of software on the platform. The current technique is to programmatically trigger a breakpoint on a secondary core when a breakpoint on a primary core is hit. This is not an ideal solution because, depending on the platform, it can take microseconds for this to take effect and you will never get reliable and consistent breakpoint synchronization that way. The only consistent solution is to have the processor incorporate a hardware synchronization unit that forces breakpoints to be synchronized on request between the cores that have a strict execution flow dependency on each other. Only this way is it possible to ensure that low-level software handshaking between several cores can be debugged and tested adequately without skewing timing relationships between the cores with every debug event (e.g., breakpoint) triggered.

10.2.6.2 Debug Handler/Application

In the application debug context the problems with multi-core debugging are a bit easier to address from a software perspective than in the JTAG system debug context. Here the challenge is to make the debugger aware of the multi-core threads and make sure it is able to trigger on thread handshaking constructs. In a homogenous multi-core processor this is simple since threading for homogenous cores are OS-specific or are based on a common well-defined library function interface such as OpenMP. Once you have heterogeneous cores this gets more challenging since there is no common standard, but the synchronization constructs tend to be specialized and unique to one of the cores involved. The challenge is going to be to find a common standard or a small set of standards so it is easier for debuggers to evolve and adapt to the needs of the developer.

Chapter Summary

This chapter summarized several facets of new low power Intel Architecture processors. First, basics on the architectural changes are detailed, including the difference between in-order and out-of-order processing. A method of evaluating power utilization is detailed using BLTK to show the benefits of compiler optimization. Third, embedded debugging is detailed, including the challenges and solutions to hardware board bringup, OS and driver debugging, and application debugging.

References

[1]   POV-Ray Version 3.1g, http://www.povray.org/ftp/pub/povray/Old-Versions/
      Official-3.1g/Linux/
[2]   BLTK Version 1.0.7, http://sourceforge.net/projects/bltk

[3]   K. Krishnan, *CPU Power Utilization on Intel® Architectures*, http:// softwarecommunity.intel.com/articles/eng/1086.htm

[4]   K. Krishnan and J. De Vega, *Power Analysis of Disk I/O Methodologies*, http:// softwarecommunity.intel.com/articles/eng/1091.htm

[5]   J. De Vega and R. Chabukswar, *Data Transfer over Wireless LAN Power Consumption Analysis*, http://www3.intel.com/cd/ids/developer/asmo-na/ eng/333927.htm

[6]   R. Chabukswar and J. De Vega, *Enabling Games for Power*, http://www3.intel.com/ cd/ids/developer/asmo-na/eng/dc/mobile/333574.htm

[7]   J. Levine, *Linkers and Loaders*, San Francisco, CA: Morgan Kaufmann, 2000, ISBN 1-55860-496-0.

[8]   http://www.moblin.org.

## CHAPTER 11

Summary, Trends, and Conclusions

The shift from designing embedded systems with single-core processors to multi-core processors is motivated by the need for greater performance within reasonable power constraints. Systems with multiple processors are not new; what is new is the widespread availability of processors consisting of multiple cores that has been made possible by continued improvements in silicon manufacturing technology. In 1993, the most up-to-date Embedded Intel® Architecture processors were built using 0.8 micron process technology and incorporated 3 million transistors in one processor core. Fifteen years later, the state-of-the-art processors are built using 45-nm process technology and incorporate over 820 million transistors in four processor cores. Chapter 2 summarized this progression of Embedded Intel Architecture processors and the key performance features added at each stage of evolution.

The cost savings attributed to consolidation of applications from multiple embedded systems to fewer multi-core processor embedded systems is another primary driver of this shift. Wireless telecommunications infrastructure, industrial control, federal, enterprise infrastructure, in-vehicle infotainment, voice and converged communications, digital security surveillance, storage, and medical are the major embedded market segments that will benefit from the performance and cost savings afforded by multi-core processors. Chapter 3 provided an overview of each of these embedded market segments and how multi-core processors can provide benefit. A critical first step in evaluating the performance benefits of multi-core processors is the use of benchmark information. A survey of existing benchmarks for evaluating multi-core processor performance was included. The survey detailed benchmarks targeting single thread, multi-core processor, and power performance and included a technique to estimate embedded application performance on new processor cores by using historical data.

Many embedded developers would like to employ Embedded Intel Architecture processors in their designs but are hampered by porting issues associated with moving from a different ISA. Moving the application requires decisions to be made with regard to 32-bit or 64-bit support, overcoming potential endianness differences, and choosing the right BIOS and OS. Fully taking advantage of multi-core processors on Intel Architecture requires sufficient OS support that is not given in embedded applications and effective software tools for multi-core software development. Chapter 4 provided details on typical porting issues and solutions to those issues in moving to Intel Architecture processors. Understanding and careful planning will determine if the features of multi-core processors can be used effectively.

Before optimizing specifically for multi-core processors, a necessary first step in extracting as much performance as possible is serial optimization. Understanding how to apply compiler technology to optimize your application increases the return on your efforts. Chapter 5 covered serial optimization and then detailed compiler usability features that help increase development efficiency.

Once serial optimization has been completed, a multi-threaded parallel optimization process is applied which begins with an analysis phase to determine the most beneficial places to implement multi-threading. Second, the design phase includes selecting the appropriate threading technology, ensuring the platform supports the necessary features for your threading, and then implementing. Debug is the third phase, which includes a standard debug cycle followed by the use of thread verification tools for diagnosing the more difficult threading-related issues. The tuning phase involves assessing the performance of the application and improving it by reducing synchronization overhead or correcting issues such as thread imbalance. The process repeats until the performance requirements have been met. Chapter 6 detailed this Threading Development Cycle, which can be applied to your multi-threading projects.

The biggest hurdle to obtaining the maximum benefit from multi-core processors is software-related; developers need easy and effective methods of obtaining performance benefit while minimizing risk. Chapters 7 and 8 detailed two case studies in a step-by-step fashion using today's threading technology to obtain benefit from multi-core processors.

With regard to multi-core processors, multi-threading is only one technique for improving performance. Chapter 9 detailed virtualization and partitioning which enables one embedded system to appear to be multiple distinct systems to different applications. This virtualization enables effective use of multi-core processors and potential cost consolidation because applications requiring the performance and consistency from separate systems can now be executed on one multi-core processor embedded system.

Chapter 10 shifted the focus to new Intel Architecture processors targeting low power devices. Topics of discussion included tips on reducing power consumption and a case study employing Battery Life Toolkit to quantify the power benefits of different design choices. The chapter concludes with a detailed discussion on embedded debugging.

11.1 Trends

Embedded application requirements will continue to evolve alongside the processing power that is being made available. Embedded applications are diverse so looking out into the future, the trends concerning application of multi-core processors are diverse; however, these trends are generalized as follows:

• Consolidation of functionality on embedded systems

• Increased data processing capabilities enabling increased workload and analysis

Chapter 3 reviewed various embedded market segments and how multi-core processors can accommodate the needs of applications in these segments. For example, in communication infrastructure, servers consolidation of applications is occurring. Functions such as SSL acceleration, intrusion detection, firewall, and VPN are being integrated into one converged security appliance. This is an example where increased performance coupled with the ability to separate system resources via partitioning enable cost savings. In DSS applications, the data processing increases afforded by multi-core processors enable scans of a greater number of video feeds and the use of more sophisticated analysis algorithms.

These application trends will have repercussions on upcoming multi-core processor design as silicon vendors create more specialized products to offer a greater degree of high performance and low cost.

11.1.1 Processor Trends

Embedded Intel Architecture Processors will continue to adopt the multi-core processors made available first in the desktop/server market segments. At the Fall 2007 Intel Developer Forum, Paul Otellini made reference to Octo Core processors; multi-core processors featuring eight processors will become available in the near future. Embedded applications that can take advantage of eight processor cores are the same ones that today employ Quad Core processors such as DSS and Medical Imaging.

A second processor trend is occurring at the lower end of the performance spectrum where characteristics such as power utilization and form factor are of greater importance. Chapter 10 discussed MIDs and the category of below 1 Watt Embedded Intel Architecture processors. These processors are evolving toward complete SOC designs where the cost and space savings afforded by integration of the memory controller, I/O controller, and graphics controller outweigh the flexibility offered by non-integrated components. In the first MID platform to incorporate the new low power Intel Architecture processor, the Intel® Atom™ processor, true SOC-type integration does not occur; instead the processor is paired with a second chip that includes the I/O and memory controller. In the successor platform, called Moorestown, the graphics and memory controller are integrated on the processor and in addition includes specialized video encode and decode acceleration. A second chip, termed the communications hub, contains the I/O components. Besides the integration planned for MIDs, other embedded Intel Architecture processors will be introduced with greater levels of integration including Tolapai SOC Accelerator and a consumer electronics media processor similar to the Intel CE 2110 Media processor. In the future, these specialized single-core[1] processors will be replaced with heterogeneous multi-core processors containing a mix of general purpose cores and acceleration components.

11.1.2 Software Challenges

The software challenges associated with multi-core processors do not go away with heterogeneous multi-core processors; in fact, they become even greater with challenges in three areas. The first challenge is programmability; requiring programmers to write code for a processor with a different ISA runs counter to two of the primary benefits of the

---

[1] Single core refers to single general purpose processor core.

IA-32 ISA: a broad software ecosystem and widespread adoption. Data communication is the second challenge and concerns the method of transferring data between processor cores in a fashion that encourages code portability. Improvements in the programming model is the third area and concerns evolution from low level thread programming to higher level models. The next three sections detail specific technologies for meeting these challenges.

11.1.2.1 Intel® QuickAssist Technology

The challenge around programmability is the desire to have one software code base that is compatible across the gamut of possible processor and accelerator combinations. For example, if I want the same application source code to work with homogenous multi-core processors and heterogeneous multi-core processors containing accelerators, how would I do so? One solution is Intel QuickAssist Technology, whose goals are as follows:

- Enable programming of accelerators to be similar to programming for the IA-32 ISA and Intel® 64 ISA

- Accelerator component appears as a functional unit through run-time and OS support

- Accelerator component shares memory with general purpose processor cores

- Run-time distributes work between cores and accelerators

An easy method of envisioning Intel QuickAssist Technology is as a set of dispatch functions that automatically take advantage of the accelerators, if present in your multi-core processor. If the accelerators are not available, the function calls will dispatch to a version that executes on the general purpose processor cores. Suppose your application code contains a call to a video processing algorithm such as H.264 decode. If your application was executing on a homogenous multi-core processor such as a dual core IA-32 ISA processor, the function call would dispatch to a multi-core optimized H.264 decode function. If your application was executing on a processor with H.264 hardware acceleration, the function call would dispatch to a version that makes use of the hardware accelerator. From an application viewpoint, the memory is shared so there is no user-visible marshaling and unmarshaling of data that would typically occur in a heterogeneous multi-core processor with distinct address ranges.

11.1.2.2 Multicore Association Communication API (MCAPI)

One effort that can help with the challenge around data communication is an industry standard communications API from Multicore Association called Multicore Association Communication API (MCAPI). Currently, there is no widely adopted message passing standard for communicating between embedded processor cores in cases where memory is not shared. The lack of a common standard makes it difficult for a broad software ecosystem to develop and makes it difficult for embedded developers to change the underlying heterogeneous multi-core processor architecture.

MCAPI is an API that defines three fundamental communication types and operations. The communication types are summarized as:

- Messages – connectionless data transfers.

- Packet channels – connected, unidirectional, arbitrary sized data transfers.

- Scalar channels – connected, unidirectional, fixed sized data transfers.

One fundamental component in MCAPI is an endpoint which defines a communication termination point. A MCAPI message would include a send endpoint and a receive endpoint in addition to the data, which specifies the source, destination, and transfer. Packet and scalar channels require the send and receive endpoint to be defined at channel initialization so that subsequent send calls require only the data to be sent and thus limit the data transfer overhead. Packet channels allow arbitrary sized data items while scalar channels limit the transfers to fixed sized elements. Scalar channels have the advantage of even lower overhead since the size of the data transfer is implicitly defined.

MCAPI does not specify requirements in terms of link level and hardware protocol and instead targets source code compatibility for applications ported from one operating environment to another. The MCAPI specification targets a wide range of multi-core applications and includes use case code for applications in the automotive control, packet processing, and image processing domains.

11.1.2.3 Software Transactional Memory

The third software challenge in future multi-core processors is an answer to the question of what will be the preferred programming APIs. Will explicit threading technologies continue to be used with the requirement that programmers know how to debug and resolve the complex threading issues detailed in Chapter 6? Or will a new programming API become prevalent?

Software transactional memory (STM) is a new model of concurrency control that attempts to overcome the inherent challenges of thread programming by offering the following benefits:

- Simplicity – programming is similar to using coarse grained locks.

- Scalability – performance is similar to using fine grained locks.

- Composition – STM regions can be composed safely and efficiently.

- Recovery – "Under the hood" access errors are automatically recovered and restarted.

An easy way of envisioning STM is to imagine the use of a coarse grained lock around a region of code that needs to be atomic. With a traditional lock-based approach all of the code and data structures would be protected and result in functional correctness, but less than optimal performance because it may be the case that the actual concurrent access that results in incorrect behavior is rare. With STM, there is a run-time system that allows parallel execution of regions marked as atomic and checks to make sure no actual parallel access to individual data elements that would create a threading issue is committed. In cases where parallel access to individual modified data occurs, the run-time system is able to rollback the operation and reprocess. This functionality gives the concurrency control its "transaction-like" property.

11.1.3 Bandwidth Challenges

Current multi-core Embedded Intel® Architecture processors use a bus architecture that requires all processor cores to share the same connection to main memory and I/O. In applications that frequently access memory and do not fit in cache, performance can be significantly impacted by this sharing. The current bus architecture will limit the ability of processors with an increasing number of cores to be used effectively. A new interconnect architecture, termed Intel® QuickPath architecture, is under development. This architecture will support point-to-point connections between processors so instead of all processor cores sharing a single communication channel (shared bus), a switch handles proper routing of multiple data communications occurring at the same time between different components such as processor cores and I/O devices. As a result, parallel communications from multiple independent processor cores are not synchronized on access to a shared bus.

Silicon process technology is allowing the creation of processors with tens and even hundreds of cores. Introduced in late 2007, the Tilera TILE64 processor has 64 general purpose processors. In late 2006, Intel announced an 80-processor core TeraFLOPS Research processor to help investigate programming and bandwidth challenges with building a processor with a large number of cores. These two processors have in common the type of interconnection layout between processors. Instead of a traditional bus architecture with all processor cores sharing the same bus and thus limiting bus access to one processor core at a time, a tiled array is employed where the processor cores possess communication channels with its nearest neighbor. The combination of a processor core and communications router is termed a *tile* in this approach. The router on the tile is capable of directing data from one processor core to another without direct processor core intervention and thus less overhead.

11.2 Conclusions

Embedded multi-core processing has arrived due to the abundance of transistors that can be fabricated on a piece of silicon and also due to the difficulty of continued performance improvement on a single thread. Embedded developers cannot rely upon the hardware to automatically provide increases in performance; it is up to the educated developer to effectively use multi-core processors and software technology to best benefit.

Appendix A

```
1  ..B1.3:               # 1              # Preds..B1.2
2          movl     $b, %eax                                  #55.2
3          movl     $a, %edx                                  #55.2
4          movl     $c, %ecx                                  #55.2
5          call     multiply_d.                               #55.2
```
Figure A1: Assembly Call of multiply_d() from main()

```
1   # -- Begin   multiply_d
2   # mark_begin;
3        .align   2,0x90
4
5   multiply_d:
6   # parameter 1: %eax
7   # parameter 2: %edx
8   # parameter 3: %ecx
9   ..B4.1:          # 1           # Preds ..B4.0
10       movl     4(%esp), %eax                             #11.1
11       movl     8(%esp), %edx                             #11.1
12       movl     12(%esp), %ecx                            #11.1
13
14  multiply_d.:                                            #
15       pushl    %edi                                      #11.1
16       pushl    %esi                                      #11.1
17       xorl     %esi, %esi                                #14.6
18                            # LOE eax edx ecx ebx ebp esi
19  ..B4.2:          # 3           # Preds ..B4.4 ..B4.1
20       movl     $-3, %edi                                 #15.8
21                            # LOE eax edx ecx ebx ebp esi edi
22  ..B4.3:          # 9           # Preds ..B4.3 ..B4.2
23       movsd    (%eax,%esi), %xmm2                        #17.18
24       mulsd    24(%edx,%edi,8), %xmm2                    #17.28
25       movsd    8(%eax,%esi), %xmm0                       #17.18
```
Figure A2: Assembly language version of multiply_d()(Continued)

```
26         mulsd      48(%edx,%edi,8), %xmm0                            #17.28
27         movsd      16(%eax,%esi), %xmm1                              #17.18
28         mulsd      72(%edx,%edi,8), %xmm1                            #17.28
29         addsd      %xmm0, %xmm2                                      #17.28
30         addsd      %xmm1, %xmm2                                      #17.28
31         movsd      %xmm2, 24(%ecx,%edi,8)                            #19.4
32         addl       $1, %edi                                         #15.18
33         jne        ..B4.3      # Prob 66%                           #15.4
34                                # LOE eax edx ecx ebx ebp esi edi
35   ..B4.4:          # 3        # Preds ..B4.3
36         addl       $24, %esi                                        #14.16
37         addl       $24, %ecx                                        #14.16
38         cmpl       $72, %esi                                        #14.2
39         jl         ..B4.2      # Prob 66%                           #14.2
40                                # LOE eax edx ecx ebx ebp esi
41   ..B4.5:          # 1        # Preds ..B4.4
42         popl       %esi                                             #23.1
43         popl       %edi                                             #23.1
44         ret                                                         #23.1
45         .align     2,0x90
46                                # LOE
47   # mark_end;
48
49
50
51   # -- End  multiply_d
```

Figure A2: (*Continued*)

Glossary

1U   Standard unit of measurement for server systems and the amount of space used vertically in a server rack. Equal to 1.75 inches.

Abstraction   General term used to describe techniques of taking advantage of a particular technology without the need to manage as many low level details as previous techniques.

Accelerator   Specialized integrated circuits used to increase processor core performance on specific applications. In some cases, the accelerator can be considered a separate processor core with its own ISA.

Address-space Compression   Virtualization challenge caused by the fact that the virtualization implementation and guest OSes each assume they are managing the entire address range.

Advanced Optimization   Compiler optimization that includes automatic vectorization, interprocedural optimization, and profile-guided optimization.

Alias Analysis   Compiler phase that determines if it is possible for two pointers to reference the same memory location.

Application   A collection of software that executes on an embedded system possibly providing operating system functionality, software drivers, and user-interface functionality.

Application Programming Interface (API)   Method of access and use for a software system. Typically this is comprised of function calls that collectively provide a service.

Asymmetric Multiprocessing (AMP)   Multi-core or multiprocessor configuration where two or more CPUs have a different, and in many cases disjoint, view of the system.

Automatic Vectorization   Compiler optimization that changes code that computes individually on scalar data elements to use instructions that compute on multiple data elements at the same time.

Automotive Infotainment   Embedded processor market segment that includes networked multimedia applications in vehicles.

Bandwidth   Rate that data is passed through a communication medium between structures.

Bare Metal   Embedded processors that do not execute an operating system to control system resources.

Basic Block   Set of instructions where if one instruction executes the others are guaranteed to execute.

Basic Input Output System (BIOS)   Program executed when a computer system is powered up and enables more complex programs such as an OS to load.

Benchmark   An application that is used to quantify performance of an embedded system. Examples include CPU2000/179.art and EEMBC Automotive/ttsprk.

Benchmark Suite   A collection of applications that is used to quantify performance of an embedded system. Examples include SPEC CPU2000 and EEMBC Automotive.

Big Endian   Ordering of bytes composing a larger value where the most significant byte is stored at the lowest storage address.

Binary Instrumentation   Modification of an executable which typically adds profiling hooks. The executable still functions as expected but now records events specific to the type of instrumentation that was added.

Binary Translation   The act of reading in machine language code and producing a new machine language code that is typically more optimized or for a different ISA.

Blade Server   A relatively thin computer that is stacked in a rack with other computers and peripherals typically located in a data center. Also referred to as "blades."

Byte Swap   Function than can be performed by an x86 instruction or sequence of code that interchanges bytes in a larger sized value such that the endian order is changed from either big endian to little endian or vice versa.

Cache   A memory area that sits logically between the processor registers and system memory. It is typically smaller in size and faster to access than main memory.

Cache Miss   A data access that is not stored in cache and must obtain the data from the next further structure in the memory hierarchy.

Cache Thrashing   Cache access behavior that results in data being brought into and flushed out of cache in quick succession, typically resulting in poor performance.

Call Graph   Shows relationships between a function (caller) and the function it calls (callee).

Central Processing Unit (CPU)   Processor which executes machine instructions and is the controlling device of a computer system.

Coarse Grain Locking   Concurrency control technique that employs synchronization around data elements where only some of the data elements have the potential to cause a threading issue. This approach may sacrifice performance as some data accesses are unnecessarily synchronized, but is typically easier to code than fine grain locking.

Commercial Off the Shelf (COTS)   Platform (motherboard, CPU, chipset) available from OEM serving as a building block for an embedded system.

Compiler Processor Target   Assumed processor for which a compiler is tuned to generate code.

Complementary Metal-oxide Semiconductor (CMOS)   Class of integrated circuit commonly used in microprocessors.

Cpuid   An instruction that returns information on the processor, including generation and stepping.

Critical Path   Portions of a computer program where any increase in time to execute the portion results in an increase in overall application execution time.

Critical Path Analysis   Method of viewing a thread profile that focuses upon the key thread transitions in a multi-threaded program.

Critical Section   A synchronization method which allows a region of code to be protected such that only one thread can execute the region of code at a time.

Data Acquisition   Embedded application in the industrial automation market segment pertaining to electronic measurement of physical phenomena.

Data Decomposition   Division based upon the data needing processing.

Data Dependency   Variable A is said to be data dependent upon variable B if the value of A is calculated based in part or whole upon the value of variable B.

Data Race   A condition where the ordering of two or more threads reading and modifying a value is not guaranteed.

Debug Build   Application built with an additional debug code, usually protected by preprocessing macros.

Debug Handler   An interrupt handler routine that stays idle until a debug event is detected on the target.

Debug Symbol Information   Additional data associated with a program that is used to assist the debugger provide its functionality.

Direct Memory Access (DMA)   Enables devices to access main memory directly instead of requesting the processor to modify the data.

Emulation   Interpretation of machine language code for one ISA that is then executed on a platform supporting a different ISA.

Endianness   Endian architecture of a respective platform, either big endian, little endian, or both (termed bi-endian).

Endianness Assumption   Code that assumes an endian architecture and may execute incorrectly if executed on an architecture with different endianness.

Event-based Sampling (EBS)   Profiling that employs hardware counters capable of recording processor activity and posting interrupts based upon counter overflow.

Execution Flow   Intel® Thread Profiler term for the execution through an application by a thread used in performing critical path analysis.

False Sharing   A performance issue in caches where the access and modification pattern of different values that happen to map to the same cache line cause the cache line to be repeatedly synchronized between two processor cores' caches.

Fine Grain Locking   Concurrency control technique that focuses on synchronization of access to individual data items. This approach to concurrency control is complicated to code, but typically offers higher performance than coarse grain locking.

Floating-point   Representation used to perform calculation on real numbers.

Flow   All of the packets associated with one Ethernet communication. Also termed packet flow.

Flow Level Parallelism   The ability of two or more packet flows to be processed in parallel due to the lack of dependencies between flows.

Flow of Control   Refers to the actual executed code paths in a program which contains control constructs such as loops and if statements.

Flow Pinning   The assigning of processing to a specific processor core of all the packets associated with a specific connection.

Functional Decomposition   Division based upon the type of work.

General Purpose Register (GPR)   Data space in a microprocessor core where values are operated upon.

Gigahertz (GHz)   1 billion cycles per second.

Global Positioning System (GPS)   A set of satellites that enable devices to detect longitudinal and latitudinal information. For example, a GPS-enabled device can help you determine your location on the Earth.

Hard Real-time   Characteristic of an embedded system where operations must meet a specific time constraint, otherwise a critical failure will result.

Hardware Abstraction Layer   Programming layer that is logically in between the underlying hardware and the software that executes on the computer. Provides a unified interface to allow the execution of the software on different types of hardware with minimal change.

Hardware Platform Bringup   Phase of an embedded project where the prototype hardware is booted and debugged to enable testing of the software components.

Heterogeneous Multi-core Processor   A processor comprised of more than one processor core where at least one core supports a different ISA than another.

High Availability   Reliability description of an embedded system typically denoted by the requirement for "Five nines" availability, which is equivalent to system downtime less than 5 min a year.

Homogeneous Multi-core Processor   A processor comprised of more than one processor core where all cores support the same ISA.

Hyper-threading Technology   Also known as simultaneous multi-threading, enables one processor to appear as two processors to the operating system and handle instruction streams from two different processes in parallel.

Hypervisor   See Virtual Machine Manager.

IA-32 Architecture   32-bit ISA that modern Intel processors support.

In-circuit Emulators (ICE)   Device that connects to an embedded processor and enables program debug. This debug approach is more expensive than more widely used approaches such as those that employ a debug handler.

Inlining   Compiler optimization where a function call is textually replaced with the body of the function call and thus reducing call and return overhead.

Instruction Level Parallelism   Parallel execution at the level of machine language instructions.

Instruction Pointer   Processor structure that contains the address of the current machine language instruction that is executing.

Instruction Scheduling   Ordering of machine language instructions selected by the compiler or assembly language writer based upon hardware constraints and program dependencies.

Instruction Set Architecture (ISA)   The commands that a processor understands and structures necessary to support the instructions, such as registers and addressing modes.

Integration   Incorporating functionality onto a processor that was performed by a different component in early implementations.

Interprocedural Optimization   Compiler optimization with the ability to track properties across functions and enables other optimizations.

Interrupt   System facility that enables a microprocessor to be signaled regarding the occurrence of an event.

Interrupt Descriptor Table (IDT)   A processor structure that contains the list of supported interrupts associated with a particular function address (interrupt handler) to call to process the interrupt.

Interrupt Vector   The offset into the IDT corresponding to an interrupt.

Joint Test Action Group (JTAG) Probe   Connector that enables access to an on-chip debug module integrated into the CPU, which allows efficient debugging of embedded systems.

Kernel Module   Code that can be loaded and unloaded into an executing kernel on demand.

Latency   The time period from start to completion of a unit of work.

Legacy Applications   General term for applications that have been extended over 10+ years and are currently still used in production. In some cases, these applications have issues (such as endianness assumptions) that prevent them from taking advantage of newer technology.

Little Endian   Ordering of bytes composing a larger value where the least significant byte is stored at the lowest storage address.

Lock   Synchronization method that permits only one thread to access a protected value at a time.

Machine Language   The commands a processor understands. These are typically made up of 1 byte to many bytes depending on the specific instruction.

Many-core Processor   A homogeneous multi-core processor consisting of cores with substantially different microarchitectures.

Megabyte (MB) 2^{20} bytes.

Megahertz (MHz)   1 million cycles per second.

Memory Bandwidth   Amount of data that can be transferred from memory to the processor core at a given time, typically expressed in terms of Megabytes per second.

Memory Hierarchy   Successive memory regions that differ in terms of access times and memory size. Generally, the memory region closest to the processor is the fastest in terms of access time and smallest in terms of size.

Memory Protection   Feature of modern operating systems that restricts memory access to the process that has been assigned use.

Memory-type Range Registers (MTRR)   System registers that specify access status for regions of physical memory.

Microarchitecture   The building block parts of a processor that implements support for the ISA.

MMXTM Instructions   Instruction set extensions focused upon parallel execution of integer operations.

Multi-core Processor   An integrated circuit that serves as the central processing unit and is comprised of more than one core.

Multiprocessor System   A computer device consisting of multiple microprocessors.

Multi-threading   Concurrent execution of multiple threads.

Mutual Exclusion   Property inherent in well-designed multi-threaded programs that prevent concurrent access to shared data.

Name Decoration   Also called name mangling, in C++. Actual function names in object file are decorated based upon function arguments to support function overloading.

Nops   Assembly instruction that means no operation. The instruction takes space in the executable but performs no action.

Out-of-order (OOO) Execution   Ability of a processor core to execute instructions not in program order through the use of special facilities to track true dependencies between instructions.

Parallelism   The execution of two or more tasks at the same time.

Partitioning   Assigning cores in a multi-core system to execution domains comprised of different operating systems.

Performance   Key characteristic of a computer system upon which it is compared to other systems.

Performance Monitoring Counters (PMC)   Resource in a processor core capable of counting the number of times specific events occur during execution.

Pipelining   Abstract concept used to describe where work is broken into several steps which enable multiple tasks to be in progress at the same time. Pipelining is applied in processors to increase processing of machine language instructions and is also a category of functional decomposition that reduces the synchronization cost while maintaining many of the benefits of concurrent execution.

Point-of-sales (POS)   Embedded device that functions as an enhanced cash register.

Power Utilization   The amount of electricity used to operate an embedded system. Good power utilization implies a low amount of power is used.

Preemptible Operating System   Type of OS that allows the interruption of the currently executing process.

Prefetching   The act of bringing data into the cache before the data is referenced. This is a performance optimization typically based upon analysis of past memory references.

Process   Operating system entity that is a schedulable task for execution on a processor core. Conceptually, a process contains an instruction pointer, register values, a stack, and a region of reserved memory that stays active throughout the life of a process.

Processor Affinity   Ability of an OS that allows tasks to have a preferred processor for execution associated with it.

Processor Target   The intended type of processor that will be used for application execution.

Producer/Consumer   Relationship between two threads where one thread's output is another thread's input.

Production Build   A build of the application that may not include debug path code but is instead optimized for execution time performance.

Profile   Data that represents the run-time behavior of an application.

Profile-guided Optimization   Multiple phase compiler optimization that monitors and records data on the execution of the application and then applies the data during a subsequent compilation.

Ray Tracing   Graphic technique where an image represented in three dimensions is transformed into two dimensions based upon what objects are visible from a specified field of view.

Real-time Response   Property of embedded systems and OSes that can guarantee handling of requests within a defined period.

Register Allocation   The assignment of values to named registers with the goal of reducing the need for loading values from memory and storing values to memory.

Ring Aliasing   Virtualization challenge caused by OSes which are designed to execute in ring 0 actually executing in ring 3 in a virtualized environment.

Runtime Context   The execution time state associated with a process, including the instruction pointer, register values, and stack.

Scalability   The degree to which an application benefits from additional processor cores; more processor cores should result in a corresponding increase in performance.

Scalar Optimization   Optimization of code that does not make use of multi-threading or partitioning.

Shared Memory   Property of systems where two or more processors access and use the same memory region.

Shared Variables   Values that are accessible by multiple threads.

Simultaneous Multi-threading   See Hyper-threading Technology.

Single-core Benchmark   Benchmark program that is a single process and single-threaded and therefore would not significantly benefit from execution on a multi-core processor.

Single Instruction Multiple Data (SIMD)   Parallel processing technique that allows one instruction to specify operations on several pieces of data.

Soft Real-time   Characteristic of an embedded system where operations must meet a specific time constraint, otherwise a failure will result. Unlike hard real-time, the failure is non-catastrophic, resulting in for example a degradation in quality of service.

Spatial Locality   Data values that are close to each other in memory tend to be accessed close together in time.

Stack   Structure that stores data and supports two operations, push (place) data on the top of the stack and pop (retrieve) data from the top of the stack.

State in Routines   Routines with values saved and used across invocations.

Static Variable   Variable defined in the lexical scope of a function that maintains value across function invocations.

Streaming Access   Data reference pattern where repeated references are occurring in some fixed sized increment or decrement from the previous reference.

Streaming SIMD Extensions (SSE)   Instruction set extensions predominately focused upon parallel execution of single precision floating point values.

Superscalar Execution   The ability of a microprocessor to execute at least two instructions at the same time and at the same stage in the instruction pipeline.

Symmetric Multiprocessing (SMP)   Multi-core or multiprocessor configuration where two or more CPUs have a similar view of the system and share the same main memory.

System-on-a-chip (SOC)   A processor that integrates CPU, a memory controller, and I/O controller onto one integrated circuit.

Task   An operating system process. Also used generally to denote a work item.

Task Switching   Operating system function where the currently executing process is temporarily halted and another process is loaded and made active for execution by the processor core.

Temporal Locality   A data value that is referenced will likely be referenced again in the near future.

Thermal Design Power (TDP)   Worst-case power dissipated by the processor while executing software under normal operating conditions.

Thread   Software, operating system entity that contains an execution context (instruction pointer and a stack).

Thread Pool   A collection of threads that are spawned before the actual work that is to be accomplished by the threads. These threads are then requested from the pool to accomplish the work and are placed back in the pool after completion. A thread pool reduces the cost of thread creation and deletion in some applications.

Thread Profile   A log of thread behavior of an application which typically includes degree of concurrency of the application, lock start and end time, and lock contention.

Threading   The coding of software to utilize threads.

Threading Development Cycle (TDC)   Process that can be employed to effectively multi-thread application code.

Throughput   The number of work items processed per unit of time.

Tile   Combination of a processor core and communications router found in a tiled multi-core architecture, such as the Intel Terascale Research Processor.

Turnaround   See Latency.

Virtual Machine   System that offers the expected functionality associated with a device, but is actually implemented on top of a lower level system. The typical example is the Java Virtual Machine, which specifies a mode of operation for a virtual processor that is subsequently emulated on a different processor architecture.

Virtual Machine Manager   Virtualization term that refers to the combination of hardware and software that enables multiple operating systems to execute on one system at the same time.

Virtual Memory   The ability of a processor and operating system to behave like there is an unlimited memory region when in fact memory is limited.

Workload Consolidation   The placement of two or more applications that executed on separate systems in the past onto the same system.

Workload Migration   The movement of applications from one system to a different system during run-time to help improve machine utilization.

X86 Processors   Generic term for all processors whose instruction set is compatible with the 8086 processor and successive generations of it.

Index

CPSIA information can be obtained at www.ICGtesting.com
Printed in the USA
BVOW051849030912

299322BV00003B/26/P